PRAISE FOR W

T0029114

"Leah Hazard conducts a searching and ~~com~~ investigation into 'the most miraculous and misunderstood organ in the human body.' . . . All but the most learned medical historians will be astonished by what Hazard reveals, both in the scope of what the womb can do and in the work it has taken, over several centuries, to produce our still-evolving body of knowledge about the organ."

—*New York Times Book Review*

"Leah Hazard approaches a fascinating topic with professional expertise and lively human sympathy."   —HILARY MANTEL

"Hazard delivers a bravura cultural history of the uterus and the politics that surround it. . . . Hazard's eye is keen, her range broad, and her tone scrupulously compassionate. . . . This is essential reading."   —*Publishers Weekly* (starred review)

"Hazard fearlessly tackles the myths, history, and science of the uterus in this new book. . . . A revelatory, straightforward, and important work."   —*Library Journal* (starred review)

"Meticulously researched and powerfully told, *Womb* is an awe-inspiring exploration of one of the most misunderstood organs of the human body. Compassionate and compelling, Leah Hazard's vital new narrative reveals the importance of understanding the uterus for body autonomy, reproductive justice, and human rights. A phenomenal book."

—ELINOR CLEGHORN, AUTHOR OF *Unwell Women*

"An erudite, compassionate, and fascinating biography of a much-maligned organ. *Womb* is sharp and political, learned and wise, and urgent and necessary. Above all else, Leah Hazard is a brilliant storyteller. I loved it."

—KATHERINE MAY, AUTHOR OF *Wintering*

"Page for page, I may not have ever learned more from a book. And I enjoyed myself throughout. Yes, *Womb* is a history book as well as a biology book, but it's also an adventure and a celebration. It's sensitive but unflinching and a very, very worthy introduction to an organ I once inhabited but can only now say I truly appreciate. I loved this book."

—ROB DELANEY,
ACTOR AND AUTHOR OF *A Heart That Works*

"Well-researched and enlightening."   —*Kirkus Reviews*

W●mb

W●mb

The Inside Story *of* Where We All Began

Leah Hazard

ecco

An Imprint of HarperCollins*Publishers*

For everyone

Some names, identities, and circumstances have been changed in order to protect the integrity and/or anonymity of the various individuals involved.

WOMB. Copyright © 2023 by Leah Hazard. All rights reserved. Printed in the United States of America. No part of this book may be used or reproduced in any manner whatsoever without written permission except in the case of brief quotations embodied in critical articles and reviews. For information, address HarperCollins Publishers, 195 Broadway, New York, NY 10007.

HarperCollins books may be purchased for educational, business, or sales promotional use. For information, please email the Special Markets Department at SPsales@harpercollins.com.

Ecco® and HarperCollins® are trademarks of HarperCollins Publishers.

A hardcover edition of this book was published in 2023 by Ecco, an imprint of HarperCollins Publishers.

FIRST ECCO PAPERBACK EDITION PUBLISHED 2024

Designed by Jennifer Chung
Watercolor background © VolodymyrSanych/shutterstock

Library of Congress Cataloging-in-Publication Data has been applied for.

ISBN 978-0-06-315763-7 (pbk.)

23 24 25 26 27 LBC 5 4 3 2 1

The body is not soiled. Is not filth to be forgiven.
The body is not an apology.

—SONYA RENEE TAYLOR,
"THE BODY IS NOT AN APOLOGY"

Contents

In Search of the Womb

Where better to learn about anatomy than a museum dedicated to the wonders of the human body?

Serendipitously, that's exactly where I find myself on a bright October morning when even the stone spires of Edinburgh seem to wink in the cool autumn sun. I'm early to meet a friend in this city with its grisly history of body snatchers and ghosts, and as I pass the imposing archway of the Royal College of Surgeons, an inscription on its threshold presents an invitation too tempting to ignore. "Hic sanitas," say the letters etched on the pavement. Here is health.

Ten years ago, I visited the Surgeons' Hall Museums with my children, oohing and aahing over the rows of "things in jars," as the gallery's brochure puts it, and at the spotlit dioramas of tailcoated doctors hunched over mannequins with gory papier-mâché wounds. Since then, I've trained and practiced as a midwife, working in labor suites, community clinics, triage units, and ante- and postnatal wards. In doing so, my fascination with anatomy has taken on a distinctly obstetric slant. The female reproductive system is my passion as well as my professional milieu—the way it functions and malfunctions, the way it brings forth life or causes death, the way it yields joy and pain in equal measure. Today, the idea for this book about the most miraculous and misunderstood organ in the human body is in the earliest stage of gestation: a flicker of inspiration; a moment charged with possibility. Today, I'm here to see the wombs.

I see the Obstetrics and Gynaecology exhibit signposted toward the back of the second floor and hurry toward it. First, though, I have to navigate the many organs deemed by the curator to be shinier and sexier to the visitor. Like a supermarket with all of its sweetest treats racked up front and center, the museum opens with a sizable showcase of military medicine. Bits of blasted skulls and amputated limbs illustrate the many ways in which men have hurt and healed each other on the battlefield. This, apparently, is glorious. I hustle through the aisles. It's not that I'm not impressed, but I'm after something a bit different today: bits of the "weaker" and "fairer" sex; organs that have seen the havoc wrought by birth and the vagaries of the female life cycle.

I move on through livers and bowels, a perforated appendix, a heart with a stab wound swishing across its gray, bloated chambers. There are stripped veins and a foot in the Vascular Surgery room; dull, staring eyes in Ophthalmology; misshapen jaws in the Oral and Maxillofacial display. Dawdling briefly in Urology, I count twenty testes and numerous penises in various stages of sickness and health. I look again at my map to make sure I haven't missed my destination: no, keep going, back and back into the depths of the museum.

Passing an impressive array of aneurysms by the rear staircase, I turn a corner and then there it is: Obstetrics and Gynaecology, the smallest section in the museum, with just four shelves of specimens. I try not to be disappointed; I stop and study each jar in turn, giving every organ the respect it deserves, wondering at the women whose bodies were flayed and fragmented in the name of science. There are thirteen uteruses—fewer than the testes around the corner, I note— some bloated with fibroids and cancers, and one with the slim white snake of a contraceptive coil still nestled in its flesh. A disembodied vulva still bears a tuft of startlingly bright ginger hair: a signal flare from the past, its meaning lost. There are no names, no personal details given apart from the briefest of diagnoses printed on cards. These organs, the seats of human life, are unsettlingly inert; the accompanying descriptions do not indicate which of these wombs have borne children, although given the fact that most of the specimens were harvested a good hundred years ago, before the advent of reliable contraception, there's every chance that almost all of them did.

As if to underline this function—or perhaps to compensate for the relative paucity of the exhibit—an eighteenth-century "Obstetrical Chair" with stiff, varnished struts has been placed in the corner. "The base," a card explains helpfully, "can be anchored to the floor," as if the birthing woman is so volcanically powerful—or, perhaps, so dangerous—that she must be tethered to the Earth, lest the force of her labor shoot her into orbit like a rocket. As a midwife, I've borne witness to this power many times—women transformed into raging demons, their bodies racked by each contraction of the uterus, their eyes on fire. These wombs suspended in formaldehyde are long dead, though, and silent. They hold their secrets quiet and close.

Two young women interrupt my reverie. Passing through Obs and Gynae, they shiver and recoil at the organs on display. "Go, uterus," deadpans one of the women to her friend as they grimace at the disembodied wombs and hurry to the next room, Otolaryngology, taking time to admire the ears and noses, and then lingering over the apparently less offensive infant limbs in the room beyond.

Something about the wombs sitting silently in their jars has been too much, too close, for these women. Scarier than the relics of the battlefield, more repugnant than diseased bowels and bladders.

Sometimes it's easier not to see, not to know. Mapping the body can unsettle as much as it empowers—awareness begets questions with uncomfortable answers. In this book, though, among these pages, we are made of sterner stuff and we journey with an open mind. We are ready to understand the uterus, and to find out where we all began. We stop. We linger. We learn what's inside the jar.

A normal uterus (and I use the word "normal" advisedly) is roughly 7 centimeters tall by 5 centimeters wide, with walls about 2.5 centimeters thick. The organ is sometimes said to resemble an upside-down pear, although in the final stages of pregnancy, a uterus can expand to the size of a watermelon. The female reproductive system is often described in culinary terms—a womb like a pear, ovaries like almonds, a fetus like a plum or a tangerine—perhaps to render the parts sweetly benign, tender morsels of sugar and spice and all things nice. This, after all, is a truth sung to us in rhyme from our earliest days and repeated by society ad nauseam: that girls are delicious and there for the tasting. From this point on, though, this book will eschew all food metaphors. We will learn that the uterus is far more than a sweetmeat or an empty vessel. We are learning, now, that the womb is a muscle. We can compare it quite accurately to a clenched fist, not only in size, but in power.

In fact, the uterus is remarkably similar in size and structure to another, far more celebrated organ: the heart. Like the heart, it is composed of three layers: in this case, there is the endometrium (an inner layer, which thickens and sloughs off each month as a menstrual period, and which nourishes both embryo and placenta in pregnancy); the myometrium, a smooth muscle layer formed of tightly woven fibers that can flex and relax, causing cramps or contractions; and the outer perimetrium, a filmy, visceral cover.

On either side of the uterus are slender tubes leading to the ovaries, where eggs are stored, and at the bottom or "neck" of the uterus is the cervix, a kind of fleshy gateway to the vagina. This is the diagram which many of us were forced to draw and label at school, although that knack seems to fade as we get older. According to surveys in 2016 and 2017 by the Eve Appeal, a gynecological health charity, many young women could not accurately name the parts of the female reproductive system.[1] Only about 50 percent of all men could identify a vagina on an anatomical illustration, and as for their ability to locate the uterus . . . the less said about that gaping cavern in the public's knowledge, the better.[2]

To make matters somewhat more complicated, the "normal" womb has infinite variations, some of which are surprisingly common and some of which are almost implausibly rare. For example, the position of the uterus within the pelvis can vary widely: the anteverted (forward-tipping) position, in which the uterus leans onto its neighbor, the bladder, is only found in 50 percent of women. The rest are evenly split between midposition (self-explanatory) and retroverted (tilted back toward the bowel). In this case, the "norm" actually only describes about half of us.

Some people, in fact, have uteruses that bear very little resemblance to those diagrams at school. There is the unicornuate womb—not, sadly, a mythical horse prancing through the pelvis, but rather a uterus that has only one side or "horn" branching off to a single tube and ovary. And my favorite of all, the bicornuate uterus, possessed by about 3 percent of all women: a roughly heart-shaped womb, with a sort of dip in the top of the organ that makes pregnancy slightly riskier but still eminently possible.

A small but significant number of women are born with two uteruses (the uterus didelphys), each of which can gestate a fetus conceived at different times, producing "twins" who are actually different ages. Some women, too, are born with no uterus at all—the extravagantly named Mayer-Rokitansky-Küster-Hauser syndrome, or MRKH—often only becoming aware of this variance when the teenage years come and go without any sign of a period. Pioneering transplant surgery now offers some of these women the promise of pregnancy, as we'll explore later.

We can see, then, that the concept of a normal uterus is, in many ways, a subjective one. The womb can be tipped or tilted, small or large, have one horn or two, or simply not be there at all. It's also important to understand that even a man can have a uterus, although the presence of said organ might come as a surprise. Consider the case of a seventy-year-old Indian man who, having fathered four children from what appeared to be a male reproductive system in full working order, began to experience nagging pain in his genitals. On presenting to his doctor, the man was found to have a kind of testicular hernia with a partially formed uterus hidden inside.[3] A similar fate awaited a thirty-seven-year-old British man who sought help for blood in his urine. Fearing a diagnosis of bladder cancer, the man was given better but no less shocking news: a long-dormant womb was having a period through his penis.[4] Thousands of miles and one year apart, these men both experienced the same anomaly: a quirk of fetal development in which the reproductive duct running down an embryo's tail end forms a combination of externally male and internally female genitalia.

Indeed, men can have wombs, and not just those men deemed biologically male at birth, but also those who affirm their maleness later in life. Some trans men—assigned female at birth, but choosing to live in alignment with their deeply felt male identity—opt for surgical removal of the uterus. Others, however, choose to retain their womb; depending on their hormonal treatment and desired lifestyle, these men may continue to have periods, or even birth a child. This unique scenario is one to which we will return later in the book.

—

WITHOUT A DOUBT, THE "NORMAL UTERUS" IS A SOCIAL construct—if, in fact, it exists at all. We know that most women have a uterus that looks and behaves a certain way: that pretty little pear, cute and compact, just like the picture we all had to draw in school. But we are also beginning to understand that for many women—and even for some men—the uterus can look different, declare itself in different ways, and do some rather unusual things.

"Go, uterus," indeed.

Womb

Uterus

IN YOUTH AND AT REST

I feel a thousand capacities spring up in me.

—Virginia Woolf, *The Waves*

What is the uterus doing when it's not preparing to have babies, gestating babies, birthing babies, or recovering from having babies? That question is seldom asked in a society that has come to value the womb primarily for its role in reproduction. In the eyes of the industrialized Western world, the uterus is only of interest when it fulfills its promise of new life—a vessel for the next generation, rather than an entity worthy of study and consideration in and of itself. The womb in its mature, fertile prime holds endless fascination for science and society alike, with every generation of researchers probing anew the double-edged dilemma of infertility and contraception, the mysterious ebb and flow of menstruation, and the apparent miracle of pregnancy and birth, from minuscule cluster of cells to bawling infant. But what's the womb doing when it's just . . . hanging out? The question seems both mundane and radical—suggesting the possibility that the uterus at rest could be worthy of examination and that, in turn, the organ may be of some intrinsic value to its owner above and beyond reproduction.

If we are to make any serious effort to explore the uterus outside the context of childbearing, then it makes sense to begin at the beginning, in infancy. It may be uncomfortable to think about the uterus of a baby girl, but before we do so, I'd ask you to sit with that discomfort for a moment and interrogate it. Why shouldn't we think about the anatomy and physiology of an organ in its neonatal state? When a female is born, her tiny uterus is simply that: an organ. Not yet fertile, not yet reproductive, not yet subject to the many ideals, taboos, and emotions we later project onto it, nor bound by the social norms and innumerate laws we will soon use to regulate and restrict its functions. This organ—smooth, pink, new, and vital—is just *there*, thrumming with the pulse of its owner, as neutral and mute as a lung or a liver. As we imagine this little womb, I'd argue that the unease we may feel says more about our society's sexualization of young women and girls than it does about the organ itself. To contemplate the infant uterus is to be a hair's breadth away from the infant vagina (which, too, is just *there*, existing, minding its own business), and in a world in which girls are sexualized and stereotyped at ever-younger ages, such thoughts can invoke fury, prurience, and shame. But here, on these pages, we are ready to look at the uterus at rest—even the infant uterus, nestled snugly in its little pelvis—with a clear, inquisitive, and untroubled eye.

As one might imagine, there are relatively few studies of the neonatal womb compared to those of the mature adult version. What few papers there are tend to comment fleetingly on the young organ's size and shape, rather than what might be going on inside it, and so we start with these simple dimensions: shaped like a tube or a spade, rather than the classic inverted teardrop of its adult form, the infant uterus may be 2.5 to 4.5 centimeters long, and approximately 1 centimeter thick.[1] In its very earliest hours after birth, the neonatal womb and its lining are still influenced to a certain extent by maternal estrogen and progesterone, but these levels tail off in the first week of life, often resulting in a moment of startling fear for which many new parents are completely and utterly unprepared: the arrival of the pseudomenses, or false period.

In my time working as a midwife on the postnatal ward, I became accustomed to new mothers approaching me at all hours of the day and night, pale and panicked, brandishing various unlikely bits of detritus from delivery—a clot saved on a pad for examination, a stray piece of suture material found in a gusset—but none provoked as much alarm as the tiny nappy streaked with pink. "My daughter's bleeding," they would exclaim, simultaneously embarrassed and concerned, and often more than a little bit disgusted.

What these women had noticed was a normal, physiological process about which—like so much of female life—nobody had warned them. Just as the mother's pregnancy hormones have caused a temporary thickening of the lining of her daughter's tiny womb, so, as those levels of inherited estrogen and progesterone diminish after birth, that little lining sloughs away and leaves the child's body in the form of what is essentially a mini-period (only without an egg, or any potential for pregnancy). A few words of explanation are often enough to reassure a new mother whose daughter has experienced this physiologically normal event, but at the same time, that conversation and our need for it are reminders that even from their earliest days on this Earth, female bodies are emblems of ignorance, fear, shock, and shame. They need not be—often the explanation is far simpler than whatever imagined horrors lie in the void easily filled by knowledge—but this is a story written long ago, and a narrative that follows women quite literally from the cradle to the grave.

-

Rather than consider the true form and function of the womb in all its messy, unpredictable, and sometimes disgusting truth, science has long preferred to imagine the nonpregnant uterus as a kind of crystal ball—unblemished and pristine—an inert object that only has meaning insofar as it forecasts the future of the fetus. In projecting its ideals about female purity and virginity onto the most female of all organs, science created a doctrine—the sterile womb paradigm—that has only recently been challenged in a meaningful way.

Like many of the theories which still dominate science in the present day, this paradigm was first outlined by a white European man; in this case, Theodor Escherich, a German-Austrian pediatrician with an extravagant mustache and a penetrating stare. Unlike most serious scientific doctrines, though, the idea of the sterile womb emerged from humble beginnings: in this case, a thick, tarry soup of meconium (in layman's terms, newborn baby poo).

From his early career in Vienna, Escherich traveled to Paris, where he attended lectures given by leading lights of the day, including neurologist Jean-Martin Charcot, whose theory of hysteria posited the female body as a dangerous site of mental and physical disease. Escherich's own fascination with the latter propelled him on to Munich, where he studied the biochemical properties of meconium passed at varying intervals after birth.[2] Malodorous though these experiments must have been, they appeared to prove an important point: that the infant gut is initially sterile, and only becomes colonized by microorganisms in the first few hours and days of life outside the womb. The womb itself was—or at least seemed to be—a completely clean environment in which the fetus grew and thrived.

This idea gained rapid acceptance among Escherich's colleagues—whether because of the rigor of his methods, or because of the doctrine's convenient mirroring of contemporary tropes about maternal virtue. In 1900, French pediatrician Henri Tissier picked up the baton and was the first to pronounce, "The fetus lives in a sterile environment,"[3] theorizing from his own experiments that the newborn gut starts off pristine until becoming colonized during transit through that notoriously treacherous passage, the vagina. Thus, the sterile womb paradigm, as it came to be called, was adopted as a neat intersection of pediatrics, obstetrics, and misogyny. To the early-twentieth-century, male-dominated scientific establishment, the idea that a fetus could only be colonized—one could even say contaminated—after contact with its mother's genitalia must have seemed like an undeniable and inevitable truth.

However, any keen student of science—or even casual observer of society—knows that truth is a shape-shifter, evolving according to the values and preoccupations of its particular place and time. The sterile womb paradigm held sway for years, but now, in these early decades of the twenty-first century, science and society have moved on enough to consider a new kind of truth, one that sees the uterus not as a crystal ball—cold and sere—but as a rich, vibrantly populated environment.

Life inside the womb, many scientists now believe, is not restricted to the nine months of gestation. Even the nonpregnant uterus—the womb at rest, the womb that's been so long ignored—may be home to a thriving microbiome: billions of native microorganisms, from bacteria and fungi to viruses and yeasts, with far-reaching influence over a woman's health, from her fertility to her immune system to her predisposition to cancer. As Dolly Parton sings, "The magic is inside you. There ain't no crystal ball."[4]

—

TO UNDERSTAND HOW THE UTERUS WENT FROM MICRO-bial desert to teeming metropolis in the popular scientific imagination, we must first return to our old friend, meconium. By the time the twentieth century clicked over into the twenty-first, new technologies had made it possible to detect microorganisms by identifying the tiniest fragments of residual genetic debris. Armed with these sophisticated tools and techniques, researchers turned their attention back to baby poo, with intriguing results: contrary to the assertions of Escherich, Tissier, and their many disciples, the germ-hunters of the new millennium found that bacteria appeared to be present in meconium excreted at or just after birth.[5] The surprising discovery wasn't so much that microbes existed in the guts of babies whose mothers were known to have infections at the time of birth. No, the finding that would soon bring microbiology, immunology, and gynecology together in the most unexpected way was the discovery that even the poo of babies born to healthy women appeared to be colonized by a diverse variety of bacterial species. Considering these infants had only ever lived in one environment—the womb—prior to birth, it stood to reason that the only place where this transformation could have occurred was the supposedly "sterile" habitat of the uterus itself.

As new methods of analysis began to yield equally novel results, scientists raced to collect and study samples from every possible substance produced in or around the uterus: test tubes, slide plates, and centrifuges in labs around the world brimmed with amniotic fluid, endometrial tissue, umbilical cord blood, and assorted fragments of placentas and their membranes, along with, of course, meconium. Study after study appeared to confirm the existence of a dizzying array of microbes within the womb, from ostensibly harmless "commensal" bacteria to nasties like streptococci and Escherichia coli (named after our friend Theodor, and commonly known as E. coli).[6,7] Results varied, and some detractors insisted that these findings were deeply flawed, with microbes only appearing to have been detected due to bacterial contamination from the research environment or the chemical solutions used in each experiment.[8]

It seemed impossible that a paradigm as deeply entrenched as that of the sterile womb could be overturned in a matter of years, and yet, as the chorus of disapproval grew stronger, so, too, did the data from research into this "new" phenomenon. In 2016, a Belgian team collecting tissue from the lining of the womb announced that, out of the 183 "sequences" or tests run on these samples, *all* of the sequences demonstrated the presence of fifteen different types of microorganisms. The team were confident enough in their results to declare them "consistent with the presence of a unique microbiota . . . residing in the endometrium of the human non-pregnant uterus." They went on to speculate modestly that "the uterine microbiota are likely to have a previously unrecognized role in uterine physiology and human reproduction."[9]

This simple but scientifically radical premise has transformed female reproductive health over the past decade, and is likely to revolutionize the way we prevent, diagnose, and treat gynecological and obstetric diseases—from fibroids to infertility, from endometriosis to preeclampsia—in years to come. To understand the massive implications of this new field of science, I went to Sydney—well, Zoomed to Sydney, given the limiting circumstances of a global pandemic at the time of writing—and spoke to a woman whose work on the uterine microbiome could enable early detection of a cancer that kills over three hundred thousand women—women like her, like me, and maybe like you, your partner, or your mother—every year.

-

AS SHE FLICKERS INTO VIEW ON MY COMPUTER SCREEN, Dr. Frances Byrne wears the pained expression of a parent desperately trying to appear professional while her child voices their own more urgent needs just out of shot. It's 8 a.m. for me in Scotland, but 7 p.m. for Frances in Australia, and I can hear her toddler crying that distinctive, late-evening wail of exhaustion, and then the hushed tones of her husband trying to settle their daughter while he corrals her into another room.

"I'm sorry about this," Frances says, but as soon as I mention that I have two girls of my own—and, pointing to the ladder at my side, I show her that I'm recording from my improvised "office space" underneath the eldest one's bunk bed—she visibly relaxes, and just like that, the ice is broken. We're no longer strangers in the formal role of interviewer and interviewee. We're now comrades-in-arms, fellow soldiers in the never-ending, guilt-laden war between maternal obligation and professional aspiration.

"You have teenagers," Frances says, "so you can tell me if it gets worse."

"No, it gets better," I reassure her. "There's light at the end of the tunnel."

Having acknowledged the fruits of our respective wombs, and the demands that our reproductive lives have placed on our existence, we move on to the matter at hand: Frances's pioneering study of the uterine microbiome, its relationship to disease and its potential to change our understanding of gynecological health. Her focus is the twisted love triangle between endometrial cancer, obesity, and the womb but, as she goes on to tell me, this focus could widen to encompass any number of pathologies and problems.

"Endometrial cancer is cancer of the lining of the uterus," she explains, "and it predominantly affects postmenopausal women. But of all the cancers that are known out there, it has the strongest relationship with obesity—more than 50 percent of all endometrial cancers can be attributed to being obese. But not every obese woman will get endometrial cancer. So the thing that we're trying to find out is how obesity promotes the development of these cancers. There's been a lot of research showing the impact of hormones, and the hormone imbalances that occur with obesity, and these can help stimulate cell growth and maybe help promote the development of cancer. But what's a relatively unexplored area is the role that the microbiome plays."

Enter Frances and her team at the University of New South Wales's School of Biotechnology and Biomolecular Sciences. Although there have already been studies of the uterine microbiomes of women with and without cancer, "they haven't really specifically looked at different populations of women," Frances explains. "But we're in a unique position to investigate that because we actually started collecting patient samples from obese and lean women with and without endometrial cancer quite a few years ago." When the two populations were compared, a key finding emerged.

"What we found," Frances says, "is that obese women tend to have a microbiome signature that's actually more similar to women that have cancer, whether they're lean or obese. And then the other finding was that all the women with cancer had lower levels of the lactobacillus species [in their wombs] compared to the controls." To clarify, lactobacillus is a probiotic (or "good" bacterium) found in live yogurt and other fermented foods such as miso and sauerkraut, and it is known to exist throughout the body quite happily, from the gut to the vagina. While other recent studies have indicated that lactobacillus may have protective qualities in the reproductive tract, potentially reducing or even preventing infection from HIV, the herpes simplex virus, gonorrhea, and bacterial vaginosis, none have conclusively identified the exact mechanism or process behind that effect.[10] Frances suggests that the prevalence of non-lactobacillus organisms could, in the future, be a major indicator of disease: "What these microbes are producing, and potentially the inflammation that they're causing in that particular environment, could be helping stimulate the growth of these [endometrial] cancers."

She's confident, too, that these strong early results aren't just the result of contamination. Not only is her team taking samples from wombs immediately after hysterectomy, keeping the environment as sterile and the procedure as quick as possible, but new techniques in detecting the genetic material of uterine microbes are far more accurate and sensitive than those used in the field's infancy just a few years ago.

All of this is well and good, you may think, but what do a few discarded wombs in Australia have to do with reproductive health for the rest of the world? Quite a lot, according to Frances. While I sip my morning coffee and the evening sun slants across the wall of Frances's room, she tells me that a definitive link between the uterine microbiome and the onset of certain diseases could lead to a new era of less invasive and more effective diagnostic tools and treatments for countless women.

"Maybe," she imagines, "you get the microbiome tested in your uterus, and if it's out of whack, or it's abnormal for you, or it changes after a certain procedure, maybe these are all things that could be tested in the future." And if, she continues, a woman is found to have a microbiome that's favorable for disease, whether because of an imbalance in lactobacilli or some other organism, then it's possible to imagine a future in which a sample of a healthier woman's microbiome is "transplanted" into the womb of the woman at risk. "I don't see why not," Frances says. "They're doing it already with fecal microbiome transplants." In those transplants—also known as FMT— prescreened, specially prepared feces from healthy donors is rectally administered to unwell recipients. Strange as it may sound, FMT has already shown promise in treating a variety of gastrointestinal disorders, such as colitis and Clostridium difficile infection.[11,12] Currently, more than three hundred trials worldwide are exploring the use of FMT to treat an even more diverse range of diseases, from anorexia to hepatitis.[13] Frances suggests that innovative procedures like microbiome transplants—fecal, endometrial, or otherwise—could reduce medicine's reliance on antibiotics which, in turn, has brought about one of the most acute threats to global health: antibiotic resistance.

"And that's a really cool thing to be thinking about," she adds, "that you're trying to harness the power of bacteria, rather than giving a treatment that just wipes everything out."

As I end our meeting, allowing Frances to tend to her daughter while mine can be heard video-chatting to her history teacher in the room next door, I look at my blank screen and sit for a moment with the enormity of what I've heard. The sterile womb paradigm: almost conclusively wrong. The "empty" crystal ball: an inner space of wondrous diversity and untold value. The future: a time, quite possibly, when our daughters will have their uterine microbiome sampled at the first sign of disease, followed by an infusion of healthy microbes to keep illness, infection, or even infertility at bay.

Admittedly, there's much we've yet to discover about this new frontier: routes to be explored and rejected while other vistas unfold before us, offering fresh promise—perhaps not to us, but to our children, or to theirs. While scientists have surveyed the microbiome in various states of disease, they have yet to forge a conclusive map of a "core" microbiome present in healthy females, and they also suspect that this "core" may vary among wombs in people of different ages and ethnicities.[14] What's more, many studies of this and other aspects of reproductive health still fail to present race-disaggregated data—a glaring omission, considering that Black and other minoritized ethnic women are disproportionately affected by certain gynecological conditions, from endometrial cancer to fibroids, and are notoriously underdiagnosed with others, such as endometriosis. Fortunately, the last two years have seen a number of researchers attempt to redress this balance, with early results showing strong evidence that Aboriginal, Black, and Hispanic/Latinx women tend to have markedly different uterine microbiomes from their white counterparts.[15,16] Knowledge, as the saying goes, is power, and more knowledge of these discrepancies has huge potential to empower people with wombs to stay well throughout their reproductive lives.

–

THE UTERUS AT REST, THEN, MAY HARDLY BE RESTING AT all. Even in the first few hours of life, it waxes and wanes with hormones before announcing itself, unbidden, with the pseudomenses' shocking streak of blood. As for the adult organ, once thought to be dormant and pure—an empty vessel onto which we could project our ideals of womanhood and virtue—science is only just beginning to decipher its many secrets. The answers to many of gynecology's questions may yet be found among the billions of tiny organisms teeming within each and every womb.

Periods

CRIMSON TIDE, LIQUID GOLD

I thought the thought only
Children and pious believe, that I was, just
Like that, no longer
A girl: the blood my summons, blot like a seal, a scarlet membership
Card slid from my innermost pocket. I was twelve and wise
Enough to be frightened.

—Leila Chatti, "Mubtadiyah"

Urban myth has it that we're never more than 6 feet from a rat, or 10 feet from a spider. These tales may disgust or even excite in the telling, but how would you feel if I suggested to you that you're never more than a few feet away from someone on their period? On the bus, queuing for your morning latte, on a factory assembly line, in the supermarket, even in a strip club, a first-class lounge, or an executive suite, women and more than a few trans men are quietly bleeding, their wombs doing what wombs have done for millennia: shedding their lining, starting anew, beginning another cycle of life with blind faith that this could be the month when fertilization finally happens.

Much has been written about the shame and stigma attached by cultures around the world to menstruating people and their blood. Scripture, literature, and oral history have documented the myriad ways in which girls and women have been seen as unclean, unholy, and borderline diabolical during menses—their blood having the power to contaminate and desecrate, to sabotage important events like hunts, harvests, and celebrations, and to render sex and female pleasure taboo. People on their periods have been—and in some places, still are—ostracized and sometimes even physically isolated from the rest of their community and the rhythms of daily life. Many books have explored menstruation's shameful history at length; this is not one of those books. If you have a uterus that menstruates, you already know about stigma and shame. If you've ever made the seemingly endless journey from classroom to corridor to lavatory with a tampon up your sleeve, or ever tied your sweater around your waist to hide the blooming stain of an unexpected bleed, or ever been chided by a teacher for sitting out PE with crippling cramps, then you know about shame. If you've ever tucked your tampon string into the gusset of your swimsuit, or ever cricked your neck to check for the telltale bulge of a pad in the seat of your jeans, you know about stigma. And if you're not a menstruator but you've ever blanched at the sight of a crimson, unflushed tissue at the bottom of your girlfriend's toilet, or ever hurried past the period products in a supermarket, or ever groaned and ostentatiously changed the channel in the middle of an advert for the latest, thinnest pads, then you, too, have absorbed shame and stigma even better than that pad in the advert absorbs its inoffensively blue synthetic blood. You don't need this book to tell you why the normal, physiological monthly function of an adult uterus has come to be seen as embarrassing, gross, and downright dangerous. You need this book to tell you what the womb is actually doing when it menstruates, what comes out, and why the blood you've been hiding (and hiding from) could change our understanding of disease, our bodies, and our lives forever.

Brace yourself, reader. It's shark week.

BEFORE WE GET INTO THE WHYS AND WHEREFORES OF the period's untapped potential, we need to go back to the drawing board to remind ourselves what a period actually is. If, like me, you didn't pay much attention in Health class—or Sex Ed, or Personal and Sexual Health and Relationships, or whatever name they've given to the Birds and the Bees these days—then your only knowledge of menstrual physiology may be a fuzzy recollection of a hormone chart labeled from Day 1 to 28 with seemingly random spikes of estrogen and progesterone in between. Ah, yes, the chart—coming back to you now? Let's briefly return to her, understand her, and then never speak of her again.

Between the ages of roughly ten and sixteen, most girls will have their first menstrual bleed, and the first day of this bleed—and of every cycle thereafter—is known as Day 1. Over the following few days, increasing amounts of estrogen act on the ovaries to help them mature one or more egg follicles, and on or around Day 14, a surge of something called luteinizing hormone (you can learn that one, then forget it) makes one of the follicles burst and release an egg into one of the slim tubes leading to the body of the womb itself. Progesterone helps to thicken the lining of the womb—the endometrium—just in case the egg gets fertilized by a sperm and needs somewhere juicy to implant, but if this doesn't happen, hormone levels plummet and the egg and its lining are eventually shed as what we think of as "period blood" on or around Day 28, which then becomes Day 1 of the following cycle. About 30 to 70 milliliters of fluid takes roughly three to seven days to leave the body, sometimes accompanied by symptoms ranging from abdominal cramps to breast tenderness, headaches, diarrhea, and anxiety—or all or none of the above—and the whole damn thing begins again.

You'll notice, perhaps, that when talking about periods, there's a lot of "roughly" and "about" and a fair amount of approximation. A person who menstruates may begin doing so at age nine or at age fifteen, their cycle may last for twenty-five days or for over a month, they may bleed lightly and painlessly for three days, or heavily and with debilitating pain for a week, or any combination or permutation thereof. Even the definition of "heavily" is subject to great debate: some sources suggest that this means having to change your pad and tampon every hour, or that it refers to bleeding through your clothes, or describes any bleeding at all that interferes with daily activity. As with so many aspects of gynecological health, science contemplates the menstrual period, shrugs, and throws up its hands with a muttered half explanation of what might be normal and what might not.

And as for the stuff that actually comes out—it's just blood, right? Women are taught from an early age how to hide it (the sweater-round-the-waist, the wadded toilet paper in the pants) and how to discard it (as quickly and discreetly as possible, whether flushing the evidence away or using pads touted for their cover-all scent and virtually noiseless wrappers). In TV and print adverts, the slim, happy woman in tight white jeans or tennis shorts is held up as a paragon of periods: she is the good menstruator, the menstruator who remains joyful, active, and clean. She is bleeding, but privately; she smiles, but silently.

We accept that period blood is a dirty, secret thing—a shameful secretion to be managed, concealed, and discarded—but what if I told you that the blood we are so keen to hide and discard is actually a precious source of biochemical information, with a uniquely personal signature that should be celebrated and explored? What if we knew that collecting and analyzing that blood could save us years of delayed diagnoses and painful investigative procedures, and what if those in charge of the government's purse strings knew that menstrual flow was actually gold dust, with the potential to cut waiting times and shave millions off the national health budget? The stuff we conceal, the light and the heavy, the scarlet-letter red and the winter-leaf brown—not the watery blue drip of adverts, but the real deal, straight from the source—could it be an embarrassment of riches?

—

BEFORE WE CAN CONSIDER THE IMPORTANCE OF PERIOD blood, we need to understand what's in it. The truth, in fact, is that only part of what flows out in a period—in some cases, less than half—is actually blood. One of the few comprehensive studies of this material found that, on average, only 36 percent of menstrual tissue is blood, with the other 64 percent comprised of a rich mixture of endometrial cells, mucus, native bacteria (that microbiome, again) and vaginal secretions.[1] And again, this information comes with a caveat that there is no "normal" or standard recipe for monthly flow; the same study found that the composition varied widely, with blood comprising as little as 1.6 percent in some women and as much as 81.7 percent in others. The study's authors didn't delve into the possible reasons for this discrepancy; we don't, for example, know with any certainty if the proportions of blood to other material vary according to age, race, or disease state. As with so many studies of women's health, new information poses more questions than it answers, with further investigation heavily dependent on funding and those who allocate it.

Let's return, though, to the blood—or, as many scientists have come to call it, the "menstrual effluent." You read that right: effluent, a word that conjures up dirt and debris. The *Cambridge Dictionary* defines the term as "liquid waste that is sent out from factories or places where sewage is dealt with, usually flowing into rivers, lakes, or the sea."[2] Anthropologist Emily Martin argues that menstruation has long been viewed by the medical establishment as simply the excretion of dead and useless tissue:

> *The descriptions [in medical texts] imply that a system has*
> *gone awry, making products of no use, not to specification,*
> *unsalable, wasted, scrap. An illustration in a widely used*
> *medical text shows menstruation as a chaotic disintegration of*
> *form, complementing the many texts that describe it as "ceasing,"*
> *"dying," "losing," "denuding," "expelling."*[3]

The adoption of the term "effluent" to describe period blood appears to dovetail neatly into the dominant narrative around menstruation: one which began with the taboos and superstitions of prehistoric times; inspired early theologians like Tertullian, who lived in the second and third centuries AD, to declare that "woman is a temple built over a sewer"; and continues to this day. You may balk at the use of the term "effluent," with all of its negative connotations, to describe what is a reproductively essential and physiologically healthy discharge. Here is yet another example of language denigrating and diminishing women's bodies: a dismissal, an insult. I'd urge caution at this knee-jerk reaction, though. Let's do a closer reading.

Effluent, in its truest sense, means simply "something that flows out." Those who use the term are reclaiming it from its pejorative connotations; they are merely describing what is, and what happens. They understand that what menstruators pass each month is not just blood, so we should not name it as such. We can call it something that flows, and in doing so, we recognize it not as dirt or waste, but simply a substance that goes from A to B. We can be neutral, we can open the door to possibility. Few people throw that door open and stride through it as enthusiastically as Dr. Christine Metz, and if she and her team have anything to do with it, they'll drag the medical establishment kicking and screaming through that door, too.

-

"Yuck."

In spite of Christine Metz's impressive titles—head of the Laboratory of Medicinal Biochemistry and professor in the Institute of Molecular Medicine at the Feinstein Institutes for Medical Research, and director for obstetric/gynecological research in the Maternal Fetal Medicine Fellowship Program at North Shore University Hospital and Long Island Jewish Medical Center—"Yuck" was the prevailing reaction of many of Christine's colleagues when she first proposed the research that's now become one of the most important projects in its field. Why the yuck factor? You'd think that a certain element of disgust gets trained out of doctors in medical school, what with all the cadavers, traumatic injuries, festering sores, and sloppy sick bowls one might encounter after a hospital rotation or two. Doctors—those paragons of impartiality and compassion—don't say "yuck." Or do they?

It turns out that when the substance to be studied is menstrual effluent, they most definitely do. Christine's ROSE (Research Out-Smarts Endometriosis) Study proposed that women would collect their monthly flow using menstrual cups or specialized pads and FedEx them to the research center, where clinicians could study certain cells in the blood to identify potential markers for endometriosis. The idea, Christine tells me on a video call from her desk on a bright February morning, was that abnormal stromal cells—the cells that help thicken the lining of the womb, and can enable the formation of a placenta in early pregnancy—could flag up a disease that usually takes an average of seven to ten years to be diagnosed, often with painful, costly investigations and surgeries along the way (a struggle to which we will return later in this book).

Christine is bright and bubbly at the start of our conversation—clearly delighted to share her work, and brimming with urgent enthusiasm—but she admits that the ROSE Study was a hard sell. Menstruation's yuck factor is entrenched, even though modern medicine appears to have no problem with studying other potentially embarrassing substances.

"It's shocking that it has not been well studied at all," Christine tells me. "As part of a recent review for the *American Journal of Obstetrics and Gynecology*, I looked at how many papers had been published on menstrual effluent. There were very few, versus, for example, semen or sperm." Later, replicating this search myself, I find the same outcome: only about 400 results for menstrual effluent, compared to over 15,000 for semen or sperm. The imbalance is clear.

The scientific community's oversight, Christine says, leaves a gaping void in women's healthcare. "We feel that menstrual effluent is a really important biological specimen that would tell us a lot more about uterine health, and much more beyond endometriosis, which is what we're focused on. For example, for infertility and fertility, we think it's just a gold mine, as well as for other problems like adenomyosis, fibroids, early detection of cancer, abnormal uterine bleeding, and dysmenorrhea [painful periods], which is a serious problem for many girls and women. But we feel that it's been a neglected biological specimen."

This neglect has deep roots in the shame and stigma surrounding menstruation, even among medical professionals who, perhaps, should know better. As mothers of daughters, Christine and I agree that our experiences in our own reproductive years, and with our children as they enter theirs, have brought this issue into sharp relief.

"I think physicians are reluctant to talk about their patients' periods with them in any kind of detail," she says. "And I know from my own experience, and my children's experience, that when you go to the gynecologist, you check a couple of boxes and they never ask you, 'How's your period? Do you have pain with your period? How long do you have pain? When do you have pain?' No one ever even asked me about the flow. I never knew what a heavy flow was until I had a daughter who had a rough period, basically, because nobody ever talked about it. And we think that's part of the yuck factor."

Sadly, Christine says, the yuck factor extended to her own colleagues when she reached out for support for the ROSE Study. "When we started promoting our study and trying to recruit women to the study, we found that most physicians were not going to help us. They were very reluctant to talk about the study with their patients. They said, 'Oh, my patients aren't going to give you menstrual effluent. No way. They wouldn't like to do that.'" The door to progress appeared to have been slammed shut before Christine and her team could even get a foot in.

The story, though, has a happy ending, and the ROSE Study is alive and thriving. Despite the initial resistance from physicians, it was the women themselves who came forward with eager determination. Not only were many women happy to participate when asked, but they were equally keen to complete the huge amounts of requisite paperwork that even Christine thought might be a deal-breaker. "Women who have been diagnosed with endometriosis fill out something called the WERF, which is a 40-page document [from the World Endometriosis Research Foundation]," she tells me. "But these women actually enjoy filling it in, they want to share their story with us and tell us what aspects they've been suffering with. We thought it would be horrible for them and nobody would ever fill it in!" Christine laughs and leans into the screen. "But we find the opposite." All those questions unasked by family doctors and gynecologists over the years were, it seemed, more than welcome, revealing a wealth of information that participants were only too keen to share. In turn, this information—along with careful analysis of sample material—has enabled the ROSE team to start releasing some pretty impressive results.

"We've published two papers so far that show the diagnostic is very good. The AUC—the 'area under the curve'—is 0.92, which is a very high number, showing that we are able to identify those with endometriosis."[4] In layman's terms, this means that by looking at menstrual effluent from women who are known to have endometriosis, the team have been able to identify characteristics in those cells that appear strongly indicative of disease. This early success isn't enough, though; ever looking ahead to greater, faster progress, Christine explains, "Now the question is, can we identify those with symptoms who haven't yet been diagnosed? And that study is going on. In the paper that we published [recently], there's a small subset of patients who think they have endometriosis and haven't been diagnosed yet." So far, she tells me, the cells in their effluent look very similar to those of the subjects who are already known to have endometriosis. "So we do believe that it will work," she says.

One might imagine that the medical establishment, however initially resistant, would greet these results with unhindered enthusiasm, but . . . can you see where this is going yet? Yes, the scientific community—like an enormous, stiff-ruddered ship—was slow to appreciate this change of direction. She tells me that the cell characteristics easily identified in menstrual effluent have, until now, been found only by doing endometrial biopsies, in which the vagina is held open and the cervix fixed in place so that a thin tube can be passed through it to collect tissue from the lining of the womb. "These are very invasive procedures," Christine explains. "And it's quite painful for women, and you wouldn't give repeat biopsies. But in fact, for one of my NIH [National Institutes of Health] grants, the comments were, 'That's ridiculous, why collect menstrual effluent when you could just collect biopsies on these women?' which is like the total opposite of what our approach is—that this should be a noninvasive tool that we're using, that women wouldn't mind providing, and wouldn't have to endure anything painful."

What's more, quite apart from the disregard for women's comfort and convenience, many colleagues objected to the time needed—about a month—to collect and grow the correct cells from samples of effluent. This apparent need for speed, Christine points out, is actually a nonsensical double standard. "We've had a lot of criticism that we have to grow these cells for the diagnostic, and they think that that's a big delay. Now, the real delay is that currently, most women wait seven to ten years to get diagnosed with endometriosis. What's a month to grow the cells?"

Christine explains all this to me with the barely concealed weariness of a woman who's grown accustomed to convincing resistant parties of the merit of her ideas. I recognize the expression. It's the fatigue of explaining self-evident truths, although in Christine's case, it's also tinged with sadness on behalf of the many women with gynecological problems who suffer needlessly for years because the medical establishment won't accept faster, better ways of diagnosing and managing their pain.

"The saddest part of this," she says, "is that many of the women with endometriosis have lost their jobs because they call in sick two days a month, or they don't get promotions, they don't have good healthcare . . . It goes hand in hand, they're not achieving their full potential in life. And they're actually suffering the consequences of it . . . So we're hoping to develop something that's less expensive than the surgical diagnostic which, you know, could run ten grand in this country, easily. And we think we can do it for a lot less than that."

And money, as we all know—whether the health system is privatized, as it is where Christine lives and works in the US, or nationalized, as it is for me in the UK—is the bottom line.

FORTUNATELY FOR ANYONE WITH A UTERUS, THERE ARE a few forward-thinking people making sure that investment in healthcare—whether it is privately or publicly funded—yields the best return possible. Candace Tingen is one of those people, and when she joins me by video from her home in Maryland, she tells me (after the requisite apology that "my children are on a walk, but they might come screaming in at any moment") why the intersection of money, menstrual effluent, and tech might just be the sweet spot gynecology has been waiting for.

"I'm a program officer at the National Institute of Child Health and Human Development," Candace explains, "so my job is to oversee a portfolio of grants that cover uterine fibroids and menstrual health disorders, and actually menstrual health in general." She says that while the scientific community may be reluctant to embrace new possibilities in this area, the general public are most definitely not. "It's a fun time," she says with no small amount of excitement. "Women at home were the first people to accept that, listen, I can look at my blood, I can look at my period, I can think about this and tell you how heavy it is. I mean, if you go on TikTok now, young women are talking about the consistency of their menstrual effluent, they are talking about clotting issues, they are talking about the color—they are talking about all of these things, because it's something that now is acceptable to discuss within the psyche of the younger generation that hasn't always hit the researchers and some of the older generations." I make a mental note to ask my fourteen-year-old to introduce me to the world of "Period Tok," if such a thing exists, in the same way that she's opened my eyes to the wonder of "K-pop Tok" (for fans of Korean pop music) and "Twilight Tok" (for obscure werewolf memes), and Candace goes on to tell me why this new generation of Tok and tech could be instrumental in making the latest advances in menstrual health accessible to the masses.

"We've been asking small businesses to pursue projects in this area, with bonus points if they're able to do so for menstrual effluent. So right now, there are already existing biological sensors for certain chemicals and certain proteins. The idea is that if one of these sensors was in a tampon, for example, it could pick up on a biomarker for disease. The sensor could either be within a tampon or a pad, or a separate container. So you could take a drop of your blood and put it on a small tab that would screen for that one biomarker and give you something as simple as a yes or no, this biomarker is present in your menstrual fluid. Or it could be connected to an app on your phone that scans across one or two or multiple biomarkers, and then sends that information to your clinician, who'd say, 'Hey, I'm seeing some findings that are a little bit troubling, or potentially troubling. Why don't you come in, let's take a look, let's talk about some of your symptoms, and let's go from there.' My big pie in the sky is this connection between a tampon and your phone and your clinician—I love that single pipeline that empowers women at home: point-of-care technologies that are not necessarily requiring clinician visits for everything."

As Candace speaks with unbridled enthusiasm, I have to admit that I love her idea, too, but I wonder if we'll see it realized in our lifetimes. When I ask her if she thinks that women of our age will benefit from these technologies, her answer is emphatic: "I do. I absolutely do." She tells me that one company, NextGen Jane (a name which conjures images of plucky girl astronauts in my mind), is well on its way to developing a "smart tampon" with the kind of connective and diagnostic potential that could empower menstruators to take a step toward better health outcomes with their own blood in their own homes. These developments are especially transformative for people who would have otherwise struggled to access gynecological care; Candace points out that "about half the counties in the United States don't actually have ob/gyns, so people will go to their primary care physicians for gynecologic health in many places, or they'll have to go to the next county over if they need specific ob/gyn care. And that's not even an ob/gyn who has a specialty in endometriosis or fibroids. So now you start to understand why there's such a big diagnostic delay. If we were in every primary care physician's office, when you just go to get your yearly checkup, you could just drop a tampon off during your cycle. If we could screen people and get them in and be able to funnel them into targeted care, that opens up the doors."

Although the system works slightly differently in the UK and many other countries, it is often the case around the world that people who menstruate do not have local access to specialist gynecology care, whether because of distance, cost, problems getting away from work, school, or caregiving responsibilities, or more complex barriers stemming from entrenched race- or gender-based discrimination. Candace and I close our conversation by rhapsodizing about a futuristic fantasy world in which dropping off a tampon, pad, or cup at your local clinic is just as routine as handing in a urine sample or swinging by for a blood draw. At forty-three, I might be menopausal before that happens, but for my daughter—TikTok-ing away in the next room—the smart tampon could prove very smart indeed.

-

BEING "SMART" ABOUT PERIODS—PREDICTING, UNDER-standing, and managing them—has preoccupied people since the dawn of humanity. In an article entitled "And Woman Created . . . ," published in the *Guardian* in 2004, Sandi Toksvig wrote:

> *Years ago, when I was studying anthropology at university, one of my female professors held up a photograph of an antler bone with 28 markings on it. "This," she said, "is alleged to be man's first attempt at a calendar." We all looked at the bone in admiration. "Tell me," she continued, "what man needs to know when 28 days have passed? I suspect that this is woman's first attempt at a calendar."*[5]

Nowadays, for many people who menstruate, keeping track of one's cycle is as simple as a tap on a phone. Period tracking apps have become commonplace in the developed Western world since Apple Health launched its first version in 2015. And why not? What used to be a matter of guesswork—or, for the more organized among us, a cryptic system of dots or circled dates on a calendar—is all in a millisecond's work for a computerized algorithm. It's not hard to understand why numerous period tracker apps have been developed in the past decade, and why these apps have been downloaded hundreds of millions of times. The appeal of never again being "caught out" by an unexpected bleed, or of being able to estimate the fertile midcycle window, is a no-brainer.

Women's interest in predicting and understanding their cycles, whether for reasons of convenience or conception, must surely be as old as menstruation itself. Imagine the first cave-dweller to feel a twist in her belly followed by a smear of blood on her inner thigh, or the nomadic tribeswoman who noticed that her bleeds had stopped just as her body grew heavy with child. Human nature is curious and truth-seeking; why, then, should there not be menstrual calendars—early period tracking apps, we could even call them—like the carefully marked antler bone shown by Sandi Toksvig's professor? Sadly, most early documentation of women's reproductive lives has been lost or neglected—or possibly, as Toksvig's lecturer suggests, misinterpreted—by historians and anthropologists, but, to anyone who's ever felt an unexpected rush of blood in their underwear and wished they'd seen it coming, the notion that early women *wouldn't* be interested in tracking their cycles seems far more unlikely than the possibility that they would.

The closest we can get to the truth of these early menstruators is by looking at the habits of Indigenous peoples whose practices may be long-held and relatively unchanged by the advent of modern technologies. For example, in his 2015 study of young women from the Suri tribe of southwestern Ethiopia, Dutch historian-anthropologist Jon Abbink writes that "the way Suri girls keep track of their menstrual periods is by counting the days on the basis of small knotted and beaded ropes, with each knot or bead representing a day, and the number of knots and beads signifying the stages in the cycle . . . They carry these ropes under their leather skirts and redo them every month on the first day of menstrual bleeding."[6] Simple, discreet, portable, and accurate: the Suri girls appear to be tracking their periods in a way that suits their needs and their available resources. History, on the whole, does not or cannot tell us how many other groups of Indigenous women have used similar methods, but it's hard to imagine that this Suri practice is completely unique.

Proof of the popularity of modern period tracking apps—many of which allow the user to track mood, sleep, pain levels, sexual activity, and more, in addition to the simple prediction of bleeding—is much easier to obtain. Two of the most popular apps, Flo and Clue, were recently estimated to have 100 million and 12 million monthly active users, respectively,[7] with the global market for period trackers estimated to be worth $50 million by 2025.[8] I conducted an admittedly informal poll of social media users, and of 593 respondents (almost exclusively in the eighteen-to-forty-five age group), 72 percent reported using a period tracker, with many users saying that they liked to predict their bleed so they could be better prepared at work, or they appreciated being able to track symptoms of endometriosis or cycle-related mood issues, or they liked knowing their fertile window so that they could either optimize or avoid conception. Stacey, a twenty-nine-year-old health visitor from Scotland, explained how the app had enhanced many areas of her life:

I've tracked periods and body temperature and have used no hormonal contraception for about seven years. I feel like I know my cycle really well: what my mood might be like, when to allow myself more leniency in my diet, when to lift heavy weights or do more cardio. When a friend tells me about a problem they're having—sore stomach, breakout, unprovoked diarrhea—the first thing I ask them is where they are in their cycle! It's amazing how many women have no idea about their bodies.[9]

For Stacey—and many other survey respondents who echoed her views—an app has provided a deeper understanding of her body's day-to-day functions (and dysfunctions).

In his assessment of the economic potential of period tracking apps, Professor Alnoor Bhimani of the London School of Economics argues that the apps' main appeal lies in their ability to sanitize an otherwise conceptually "dirty" process. "Importantly," he writes, "calculations centred on menstrual blood can provide a clean representation of information about a woman's body. Quantification disinfects the dirt that inhabits the reality of the object being quantified . . . Data sanitises what may have come to be considered as unclean."[10]

It's provocative to suggest that women have so internalized the entrenched societal stigma around menstruation that they've turned to apps for "cleansing," but many of the respondents to my survey did appreciate the order that these apps had brought to their lives, often using words like "manage," "predict," "plan," and—most frequently—"control." Caoilin, a twenty-seven-year-old collections consultant from Ireland, described her relief at discovering an app which could track and predict her often irregular periods: "I finally felt somewhat in control, and like I understood my body for the first time."[11] For users like Caoilin, technology has made the messy business of having a period tidier, bringing order to what is otherwise a chaotic or un-manageable experience. The woman—and not her uterus—is finally in control.

This tech-enabled autonomy is not without its drawbacks, though. In a very modern demonstration of the old adage that "you don't get something for nothing," period tracking apps are only as accurate as the data entered by each menstruator. Bhimani describes users as "prosumers," pointing out that the prosumer must surrender personal data in order to receive algorithmically calculated information, and the more one gives, the more one gets. A person who religiously en-ters the onset, duration, and heaviness of their period, along with any associated symptoms at that time and throughout the month, will glean much more from the app in return than a user who only logs on every now and again. The apps themselves reflect and encourage this relationship; detailed entries to Clue, for example, are met with a pop-up reassurance that "Clue is getting smarter."

When apps get smarter, it means that they're storing and analyz-ing more information about the user and her uterus, but with many apps monetizing this transaction by charging higher fees for more en-hanced features and content, the commodification of the womb and its functions has raised some serious ethical questions. In America, period trackers aren't required to comply with the Health Insurance

Portability and Accountability Act (HIPAA), far-reaching privacy legislation that governs the use of personal data in health-related settings, and in Europe, tighter controls around collection and sharing of app-collected data have only recently been articulated in 2018 by General Data Protection Regulation (GDPR). These closer controls haven't come out of nowhere: in 2019, an investigation by UK-based Privacy International found that apps MIA Fem and Maya had been sharing users' personal data with Facebook without adequate consent. Around the same time, Flo settled a Federal Trade Commission complaint claiming that the company had misled users and unlawfully shared information with Facebook and Google.[12] Even now, users may be unaware of the way their data is used and collected. Some, like Stacey, may not mind; she notes: "All my 100 devices are linked together and I'm probably with the generation of people who don't think twice about personal information-sharing. As far as I'm aware, [my tracker] has an easy-to-read privacy policy, but I can't say I've thought much about it." Others seeking to decipher the intricacies of data collection may be mystified by the apps' sometimes cryptic explanations; for example, a post from Clue's official Twitter account states, "Everything you track in Clue is securely stored in our backend."[13] While the statement's clear intent is to reassure users, that reassurance depends on whether users actually understand and trust the concept of secure data storage, or can identify an app's "backend."

For clarity, I decided to go straight to the horse's mouth, or in this case, the app's creator. Ida Tin, the cofounder of Clue, spoke to me from her Berlin apartment and, after the requisite small talk about our conflicting domestic and professional demands during the pandemic, she addressed the dilemma at the heart of the period tracking industry.

"Data privacy is super central," she says. "We ask people to share these periods and data. And we think a lot about how we honor that every day, and how we build trust." Responding to some of the less savory uses of consumer data, Ida concedes, "It's a huge problem in the app economy . . . and if people knew how [some] of the companies made money, they probably wouldn't want to use the product. So people have concerns about privacy, and I think, with good reason. There are many players that do, in my view, unethical things with data." In fact, she says, she wonders whether it is ethical to use period-related data at all. "There is a sort of big question around, you know, should technology be involved at all in this intimate part of life?" Clue and its competitors may be "getting smarter," but at what cost? Ida acknowledges that the technology and its use or misuse of data may be problematically opaque for the people who rely on it: "I think it would be great if there was some sort of certification, almost like 'good data practice companies' or something to that effect, because it is literally impossible for the average user to navigate how people treat data." There, in all its mystical glory, lies the "backend."

While there's no denying that multimillion-dollar apps like Clue must, at some point, be geared toward profit, Ida is keen to point out that her company's collection and analysis of data has already yielded important information about women's health. Symptoms such as unusual pain or bleeding are, perhaps, more likely to trigger cause for concern when charted by an app. "We already have so many stories from our own user base," she says, "but people say that they, in fact, did discover cancer early, or they did see that they had a pregnancy outside the womb or other things that would have been life-threatening." When seen on a larger scale, there may be value in using period trackers as experiments with unprecedented scope and sample size, offering researchers the opportunity to detect trends in reproductive health and disease among a population of millions.

"We already built an algorithm not to diagnose but to see patterns for polycystic ovarian syndrome, that we built together with a great researcher out of Boston University," Ida points out. "It's not super active in the app right now, but that could become a much bigger thing. And we could probably do it for endometriosis as well, or other things we still haven't seen." Ultimately, she says, she envisions a future in which period tracking apps can help people manage and understand their reproductive health, and to link this understanding seamlessly with access to appropriate medical care.

"I think there's something about having a longer-term data set that is really powerful, that we haven't quite leveraged on a consumer level yet . . . It's not like, I go to my health practitioner and say, 'Look, here's my full view [of data], you know, here's my landscape. Look at this, where I'm heading.' And that's something I feel is very compelling— if we could each have a sense of looking into this landscape of our health. So I think that's the promise. If we could help people really navigate through a whole life of health, that would be amazing."

Ultimately, it remains to be seen whether period trackers will be a force for good—offering users enhanced understanding of all of the womb's functions, from fertility and conception to disease—or whether the drive for profit will overtake and corrupt any altruistic motives. As with so many other areas of gynecological health, these apps are woefully under-researched; only one review has looked comprehensively at the evidence around their use, and of the 654 items of literature included in the review, a scant 18 papers were robust enough to meet the researchers' criteria for inclusion. Only one firm conclusion could be drawn: "There is a lack of critical debate and engagement in the development, evaluation, usage, and regulation of fertility and menstruation apps." With monumental understatement, the reviewers close by saying, "The paucity of evidence-based research and absence of fertility, health professionals, and users in studies is raised."[14] The trackers' backends are growing more expansive by the minute, ballooning with the seemingly endless data supplied by eager users, but until research catches up with the industry's progress, the information we have about the apps themselves will remain frustratingly limited.

–

BETWEEN TRACKING ONE'S PERIOD, CHOOSING AND PURchasing period products, managing symptoms like pain and fatigue, and the general emotional energy required to do all of these things while presenting a smiling, functional, socially acceptable face to the world, the womb's most regular function takes up an awful lot of headspace for the average menstruator. That's before we even factor in the financial costs: as much as £5,000 for period products over a lifetime (assuming you can even afford them, as many people can't), according to one source,[15] as well as the broader economic cost of an estimated five million sick days per year due to menstrual symptoms in the UK alone.[16] Then there is the physical toll of bleeding for one week out of every four: an increased risk of anemia, especially for those people unfortunate enough to bleed longer or more heavily; and a higher chance of certain gynecological cancers in line with the number of periods in one's lifetime. Why—apart from the obvious reproductive reasons—do we put up with it?

The answer—at least according to a vocal and rapidly growing minority over the last fifty years—is that we don't necessarily have to. "Menstruation has become an elective bodily process," promises a recent article in the *Atlantic*, with the tantalizing headline, "No one has to get their period anymore."[17] One enterprising gynecologist has even created the hashtag #PeriodsOptional to hammer the point home to social media–savvy millennials. Can it be true? Is there really a way to unshackle ourselves from a life of monthly painkiller-popping? Is the tampon-clutching walk of shame a thing of the past? Will we never again utter a string of creative expletives when our period tracker cheerily reminds us "1 DAY UNTIL NEXT PERIOD," or when Aunt Flo decides to turn up—as is her wont—at the most inconvenient / embarrassing / unexpected time, like during a long-haul flight, a job interview, or sex with a new partner? Every menstruator has their own horror story: a common narrative of frustration and dogged coping. Unless you are among the small number of womb-owners who believe that "moontime" is a monthly manifestation of the sacred female divine, you may hear the words "Periods optional" and reply with the words "Sign me up."

At present, the only way to suppress menstruation, either short- or long-term, is by using synthetic hormones. Anyone taking the oral contraceptive Pill will be familiar with the process: a "real" pill (containing estrogen and progesterone, or in some cases, just progesterone) is taken for most of the month, followed by a placebo pill (with no hormones) for seven or fewer days, during which a light and relatively painless bleed occurs. This last week of bleeding mimics the rhythm of an unmedicated menstrual cycle. Indeed, many Pill users call this phase a "period" in spite of (or perhaps being unaware of) the fact that it is simply a breakthrough bleed: the lining of the womb is shed only because the synthetic hormones keeping it thick have been temporarily withdrawn, and not because any kind of ovulation or other innate physiological process has occurred.

In fact, there is no clinical reason that a pharmacologically controlled cycle needs to mimic exactly the rhythms of unmedicated menstruation. The original designers of the Pill didn't decide to include the seven-day placebo "break," or bleeding phase, for health reasons; rather, some clinicians believed that this mock-period would make the Pill seem more natural, and therefore more acceptable, to the women for whom it was designed. Other sources suggest that the rationale behind the "break" was an appeal to religious leaders.

Susan Walker, a senior lecturer in sexual health at Anglia Ruskin

University, recalls, "I attended a lecture given by Carl Djerassi, the 'father of the Pill.' He remarked that the seven-day break, and resultant withdrawal bleed, was designed into the Pill in the late fifties in an attempt to persuade the Vatican to accept the new form of contraception, as an extension of the natural menstrual cycle."[18] From the outset, the use of a drug with the promise to liberate women from the burden of unwanted pregnancy was, in fact, created by one man (Djerassi) as an indirect appeal for the approval of another (the Pope). In the middle of the debate—voiceless but no less subject to the whims and desires of these powerful men—was the womb.

Pharmaceutical companies were—and in some cases, still are—keen to sell women this idea of the Pill as something both natural and liberating, allowing users to enjoy exciting new freedoms while working in harmony with the body. "On their websites and in print campaigns," writes sociologist Katie Ann Hasson, "the companies introduced potential users to menstrual suppression through images of an idealized lifestyle achievable with menstrual suppression and information that preempted questions about what is normal, natural, and safe. They presented detailed explanations of the menstrual cycle and introduced women to the 'pill period'—a new term for the 'scheduled' bleeding that occurs when women are taking cyclic hormonal birth control." Hasson argues that continued use of the phrase "pill period" by pharmaceutical companies is deceptive and undermining; she writes that some of the narratives promoted by Pill manufacturers today still "work to revise—and perhaps even delegitimize—women's knowledge and experiences of menstruation."[19] In short, language matters: calling a synthetic, drug-induced bleed a period—or implying that it is equivalent to one in any way—is a negation of a woman's fundamental right to know and understand the functions of her womb, her body, and herself. Hasson points to calls from the clinical community for the use of more accurate terminology such as "withdrawal bleeding," "bleeding days," or just "bleeding": more honest representations of what a pharmacologically controlled uterus is actually doing.

Of course, as has so often happened in history, women and people with wombs have been told how to manage their bodies, have weighed the perceived risks of bending or breaking those instructions, and have finally cobbled together their own habits in the interest of reproductive autonomy. Djerassi and his team may have designed the three-weeks-on, one-week-off model, with the "pill period" as an integral part of the contraceptive regime, but almost since its inception, users have experimented with delaying or completely suppressing this breakthrough bleed by running one or more packs of "real" pills together, skipping the placebo phase altogether. Recent research has proven that not only is this regime acceptable and even preferable to many menstruators, but it is—perhaps most important—safe. In 2014, a comprehensive review of twelve randomized controlled trials comparing the traditional twenty-eight-day Pill cycle with continuous hormonal suppression of bleeding found no difference in safety or contraceptive effectiveness.[20] In 2019, the UK's Faculty of Sexual and Reproductive Healthcare (a branch of the Royal College of Obstetricians and Gynaecologists) issued a guideline stating conclusively that "there is no health benefit from the seven-day hormone-free interval . . . Women can avoid monthly bleeding and symptoms that come with it by running pill packets together so that they take fewer (or no) breaks."[21] The verdict is in: periods—in some sense, depending on how you define them, for certain people at certain times—are indeed optional.

—

"IF THOSE WITHOUT UTERI HAD TO BLEED ONCE A month, they would have turned it off ages ago," says Dr. Sophia Yen, creator of the hashtag #PeriodsOptional. We're video-calling at what is a normal time for me in Scotland, and a crack-of-dawn time for her in California, but her energy radiates through my screen in spite of the early hour. "For so long, we with uteri have sucked it up. We need," she says with steely conviction, "to stop sucking it up."

Dr. Yen is an associate clinical professor at Stanford University, CEO and cofounder of mail-order contraceptive company Pandia Health, and mother of two girls. She's the kind of woman often glibly described as a "powerhouse"—accomplished, ambitious, and determined in her mission to provide continuous birth control as safely and efficiently as possible to as many people as possible—but that epithet doesn't do justice to the depth of Sophia's feeling, or the less glamorous origins of her crusade. The bright, bubbly woman inviting us all to "switch off" our periods was once, she explains, an anxious young premed student whose own menstrual struggles would change the course of her life.

"I was at MIT and all of a sudden, in my biochem final, I bled. And I was like, you know, do I run to the bathroom, or do I finish the exam? And, being a premed, I finished the exam, but was I a little distracted? Yes, I was absolutely a little distracted in the middle of a final. And then I looked to my left, and I looked to my right—to people without uteri—and they're like, boop-boop-boop [here, Sophia mimes nonchalantly writing on an exam paper, untroubled by sudden menses], and I realized that never in their life have they been hit by random blood in the middle of an exam. And now, looking back, fifty percent of that class had a uterus, and one in four was bleeding at that moment." Sophia lets that statistic sink in for a minute, with all of the potential pain and inconvenience of having a period during an important event, before she concludes, "What a waste."

And so #PeriodsOptional—or at least, the seed of the idea—was born. As Sophia progressed through her training and subsequent career as a gynecologist, the research around continuous contraception only reinforced her gut feeling that menstruation could and should, for most people with a uterus, be a thing of the past. She tells me that even those of us who do have periods are having far more of them than is normal or healthy, referring to an earlier time when women spent their childbearing years either pregnant or lactating.

"We've moved to the state of incessant menstruation, whereas the natural state was constantly pregnant or breastfeeding. How many periods do you have when you're constantly pregnant or breastfeeding? Zero. So we used to have one hundred periods [in a lifetime]. And now we're having three hundred fifty to four hundred: three and a half times what is, quote, 'normal.' Not just ten percent or thirty percent more, but three and a half times more, or two hundred fifty to three hundred more periods than we need. It's just unnatural to have our hormones going up and down, up and down every month. It's unnatural for us to be bleeding every single month."

The midwife in me feels compelled at this point to flag the fact that one can indeed still have a period and get pregnant even while exclusively breastfeeding, so the number of periods one might have under those circumstances would not necessarily be zero. What's more, this idealized "simpler" time in early human history was not only characterized by fewer periods but also by other equally "natural" but far less desirable events such as death in childbirth. Nevertheless, Sophia is adamant that the optimal female life span is characterized by markedly less menstruation. She tells me that research indicates a possible correlation between fewer periods and a decreased chance of certain gynecological cancers, and the good news, as she sees it, is that "there's a way to hack that—to get closer to the natural one hundred periods in our lives, as opposed to the unnatural three hundred fifty to four hundred."

Listening to Sophia's pitch, it's easy to feel the lure of a period-free life. Why bleed every month if it wasn't what nature intended, and especially if it might not even be good for us? Why should you have to worry about the stain on your jeans, or the long walk to the toilet, in the middle of an exam or any other important event? Why not just put your period on mute? In fact, why not give our daughters the head start we never had by switching their periods off just as soon as possible after they start?

"Wouldn't it be nice," Sophia asks me, "if your teenager's hormones weren't as raging, and they were kind of . . . smoothed out? And she wasn't hit with random blood one week out of four when she's clearly not trying to reproduce? I think, you know, with the Silicon Valley, Tiger Mom mentality, if your daughter has a one-in-four chance of being hit with blood during the SATs, during her finals, during an exam, during a debate—*my* daughter will have absolute control." Sophia's recurrent use of the phrase "hit with blood" posits menstruation as a kind of assault, one from which you would gladly protect your child, if only you could.

She goes on to explain why being period-free makes teenage girls not only more empowered, but potentially even more intelligent and capable than their peers, citing a study of children with and without the kind of iron-deficiency anemia that can often be caused by heavy menstruation. The anemic children "showed lower IQ," Sophia explains. "However, when they gave them back their iron, their math scores went up and their IQs went up. So maybe those of us with uteri who are bleeding one week out of four are functioning with lower oxygen running through our heads not only for those critical math permutations but also for sports, and just for breathing and living. As an adolescent medical specialist, I know that the hemoglobin in boys goes up during puberty, and the hemoglobin in girls goes down." Thus, in Sophia's eyes, nonmenstruating teens "absolutely have an edge academically."

I think of my own girls, who (thankfully) seem to be thriving at school, and of my younger self, having toiled through years of academia at various levels despite painful, heavy bleeds, and I wonder if we could have done even better without the distraction—even, according to Sophia, the handicap—of menstruation. I've tried to feed my children's bodies and minds with all things healthy and wholesome as they navigate the early twists of life's journey; should I also have offered a steady diet of contraceptive pills to level off their moods and give them an academic edge? While my daughters and I thrive regardless—at times, in spite—of our wombs, I'm still not sure whether the answer to that loaded question is yes.

When I ask whether there might be any dangers to treating girls with continuous synthetic hormones at such a formative stage, Sophia's only caveat is that it's worth waiting roughly two years after the onset of menstruation to avoid interfering with a natural growth spurt that often occurs at that time. However, when I ask her more specifically whether there might be other, possibly more insidious effects on a young person's mental and emotional development—knowing, as we do, that estrogen and progesterone can have a deep impact on mood, libido, and one's general sense of self—her reply is more equivocal. She says, "We certainly know that there's cognitive development that goes on [after adult height is reached]." As to the potential effect of hormonal contraception on this process, Sophia concedes, "Maybe there's some cognitive effect, but we don't know if it's good or bad. It could go either way. Right?" She adds that her company's average customer is twenty-five years old—well past the developmental challenges of adolescence—and that "[we're] not saying everyone should go periods optional. I'm always about choice."

In spite of these assurances, the querying note of that last "Right?" lingers in my mind long after our interview has finished. I have to make lunch for my younger daughter in between home-schooling sessions, and as I set her pasta on the table, I wonder whether there should be a small white Pill on the side of the plate. She has another two hours of classes that afternoon; she, like me, is feeling the screen fatigue, but her brain is obliged to keep churning on. I could start something now that will smooth the rough seas of her teenage years; help her navigate the exams and the upsets with clear-headed equanimity. I could give her the period-free life I never had: blissful ignorance of her uterus's monthly function, and the freedom to excel beyond her womb-bound peers. What loving mother wouldn't want her child to escape being "hit by blood" . . . right?

-

"I COULD NOT AGREE LESS," SAYS SARAH HILL, A PRO-fessor of evolutionary social psychology at Texas Christian University in Fort Worth and author of *How the Pill Changes Everything: Your Brain on Birth Control.* She's sitting cross-legged on her bed during our video call like we're slumber-party besties, but the fury in her voice tells me that she means business. As one of the foremost experts on the cognitive effects of hormonal birth control (and another mother of teenage girls), Sarah has a few things to say about putting young women on the Pill for noncontraceptive reasons.

"There are two big elements of that [idea] that I find really problematic," she begins, warming to her theme. "The first one is this really androcentric view of what it means to be successful and to compete. And the idea that competing as a woman with natural cycles is somehow a disadvantage, to me, is sexist and completely wrong-headed. I can't think of anything I disagree with more. I think that the idea that being male is the standard that we should be judging ourselves against—and that that's somehow the pinnacle of success—is bullshit." Just to make sure her argument is clear, she adds,

"I think that sends a terrible message to women. And I absolutely disagree." I think of the young Sophia Yen, bleeding in her biochem exam while the "people without uteri," as she called them, breeze through their answers, and I wonder—in fact, I think I know—how she would respond.

Quite apart from any misogynistic undertones, though, Sarah argues that using continuous contraception to suppress menstruation in teenage girls can have serious, irreversible cognitive effects. "The second element I find problematic," she says, "is that this shows absolute, reckless disregard to brain development. It's irresponsible. The brain isn't done developing until you're in your midtwenties. But, you know, even before that, there's a very critical period of brain development that's going on during adolescence, up until you're about nineteen or so, when things are rapidly changing within the brain. And the hormones that are responsible for postpubertal brain changes are your sex hormones, and the idea that you would disrupt this successful mode of brain development that's been happening for millions of years, and just sort of decide that we're going to change a person's hormonal profile with no research on what the long-term consequences are, is reckless. And especially for something as sort of, you know, *small* as turning off your period."

Sarah has hard data to back up this point: in 2020, she and her colleagues published a study comparing cognitive performance in college-age women taking hormonal contraception with that of their naturally cycling counterparts.[22] The women taking hormones spent less time and performed more poorly on both simple and complex cognitive tasks: results which directly contradict any theoretical association between physiological cycling and suboptimal performance, and which, the authors suggest, "contribute to a growing body of research examining the unintended implications of HC [hormonal contraception] use on cognition, learning and memory." In light of this conflicting evidence, it seems that the long-term consequences of altering the body and mind with hormonal contraception during a formative developmental phase have yet to be determined.

Just as #PeriodsOptional has become a battle cry for Sophia, so the science against that idea has become the central tenet of Sarah's awareness-raising crusade over the past few years. Her book explains in detail how hormonal birth control works by tricking the body into a kind of continuous luteal phase: the part of the menstrual cycle that occurs between ovulation and the onset of bleeding, when the lining of the womb thickens itself to prepare for the implantation of an embryo.

"The daily dose of hormones that you've got in hormonal birth control has a greater presence of the synthetic progesterone, or progestins, relative to estrogen," Sarah explains. "So it's kind of mimicking a synthetic, nonstop luteal phase. And that generally isn't when women feel their best. This, of course, isn't true for all women. The one thing that I try to always be really careful of when I'm having conversations about this, is that the way women respond to hormonal birth control is incredibly idiosyncratic. And the research isn't in a place where we can say, all women will respond this way . . . because we just don't know. Some women feel better on it, some women feel worse, but you do eliminate, for many women, some of the best-feeling points in the cycle when you suppress ovulation." In short, some girls' moods may be "smoother" with synthetic hormones, but such a regime may be smoothing out the good, as well as the bad, mood fluctuations of a nonmedicated or "natural" cycle. Once again, science hasn't caught up with conjecture: there is a paucity of hard data regarding the effects, cognitive or otherwise, of menstrual suppression starting from the years just after menarche.

There are, Sarah concedes, some young people who might benefit from continuous suppression of menstruation. For example, current guidance suggests that oral contraceptives may be used to control or manage the often debilitating symptoms of endometriosis, polycystic ovarian syndrome, and painful or heavy periods.[23] Sarah says, "I understand that for some women, monthly fluctuations can be intolerable, and in those cases, it might not be a reckless decision to put an adolescent girl on birth control pills if they allow her to feel sort of maximally at home in her own body, and that is the case for some, but it's certainly not the case for all."

Feeling maximally at home in one's body is undoubtedly a state to be much desired, but menstruation makes that goal harder for some to achieve than for others. Of note, hormonal contraception—whether achieved by continuous administration of pills, or by long-acting methods such as hormonal implants and devices—is a popular method of suppressing periods in girls and women with some intellectual and physical special needs.[24,25] Although "disability" is a broad term of arguably limited helpfulness, and many disabled people cope as well as their able peers do with menstruation, some find periods and the associated personal hygiene tasks distressing or difficult. Self-described "mad queer autistic disabled woman" writes in her blog, *crippledscholar*, that the needs of this population can also be at odds with those of their caregivers. She argues that "a lot of it is about control and often menstrual cessation in order to make the menstruating person more convenient for a care giver."[26]

This theory is echoed by the authors of a systematic review of the subject, published in 2019: "It is likely that disabled people face layers of discrimination when they are menstruating," they write, noting that menstrual care was widely seen as problematic and challenging by caregivers of people with intellectual disabilities. In fact, among institution-based, professional caregivers, "menstrual care was the second most disliked aspect for residential staff (after giving enemas), and the most disliked aspect for day unit staff (who do not give enemas)."[27] The message could not be clearer: for some caregivers, only the handling of actual feces is less palatable than managing the mess of menstruation. Periods may indeed be optional for disabled people in these environments, but for whose comfort and benefit? This question must be answered to ensure that menstruation is managed and supported in an ethical way for those who may already be vulnerable.

There are many more people, able-bodied or not, for whom a period and its attendant cognitive changes are unwelcome; catastrophic, even. For those trying to conceive, sadness and disappointment flow with each bleed. For some trans men, periods are a physiological reminder of an assigned identity they'd rather leave behind. And for many people with a uterus, regardless of identity or agenda, menstruation brings pain, upset, financial hardship, and yes, even in today's day and age, shame.

Equally, for many women, a monthly period is a welcome reminder of the womb's existence. That bleed may be a relief—proof that any unwanted pregnancy has been avoided, and proof of one's bodily autonomy. The oft-reported surge of libido and creativity midcycle may be cherished—a happy trade-off for the dip in mood that often comes later in the month. There may be menstruators for whom a period is, indeed, an integral part of "feeling maximally at home."

However one feels about the uterus's monthly wave hello, what's becoming increasingly clear is that the flow itself—that "effluent" so highly prized by Christine Metz, Candace Tingen, and their peers—is a valuable window onto reproductive health; a map, perhaps, of the landscape so keenly surveyed by Ida Tin and the millions of menstruators who track their periods. In a review of the National Institutes of Health's recent meeting on "Menstruation and Society," the authors argue that "reproductive health in particular has been hampered by a lack of understanding of basic uterine and menstrual physiology . . .

Enhancing our understanding of the underlying phenomena involved in menstruation, abnormal uterine bleeding, and other menstrual-related disorders will move us closer to the goal of personalized care. Furthermore, a deeper mechanistic understanding of menstruation—a fast, scarless healing process in healthy individuals—will likely yield insights into a myriad of other diseases involving regulation of vascular function locally and systemically."[28]

Many of you reading this may, as I did, have a "whoa" moment here. Stripping away the loaded gender and identity politics we often superimpose on this basic physiological process, should we simply be in awe of periods as "a fast, scarless healing process"? If we look at the monthly shedding of endometrium as a spontaneously created wound, and the regeneration of that lining as a kind of constant, cyclical healing process, occurring spontaneously every month without effort or intervention, then is menstruation actually kind of phenomenal? Children are often taught how cool it is that a starfish can regrow any lost arms; are we directing their admiration to the wrong species? Should we be telling them about the powers of the menstruating womb, and should we expect their eyes to grow saucer wide in admiration?

Again, I think of my daughters, and I imagine their ridicule if I were to try to persuade them that their periods were nothing short of miraculous. We have always been pragmatically unfussy in our house—not for us the "moontime" ceremonies and rite-of-passage rituals cherished by other, more celestially minded families—and I can't see my girls welcoming a new, enthusiastic reframing of our monthly hassles. Perhaps, though, if I told them that the effluent they hide, flush, or wrap could hold the key to their reproductive health, and if I told them that its analysis could help them avoid dangerous delays and unpleasant investigations in the future—perhaps, then, they would feel a grudging appreciation for a process which, at this point in their lives, seems to hold little purpose. In the meantime, for my daughters, for me, and for most of us who menstruate, periods remain a source of consternation and contradiction. Whether tracked on beaded belts or smartphones, whether worshipped or suppressed, dreaded or encouraged, this monthly reminder of the uterus's existence will, it seems, always evoke grief and wonder in equal measure. The one thing for sure—like death and taxes, as the old saying goes—is that menstruation happens and will continue to happen for approximately half of the world's population, from the first false period of infancy to the final farewell of menopause. The best we can do is learn from the flow in the meantime—mine that seam of gold and map the landscape around it—to enable us to be healthier, happier, and more maximally at home in our bodies.

Conception

MACHO MYTHS AND HIDDEN CRYPTS

Almighty God created sexual desire in ten parts;
then he gave nine parts to women and one to men.

—ALI IBN ABI TALIB

You know the famous scene in *When Harry Met Sally*: Sally, played by Meg Ryan, wants to show Harry how deceptively easy it is for women to fake orgasms. When Harry claims to be able to tell the difference, Sally proceeds to prove her point by simulating a full-blown, wildly vocal climax, complete with flushed cheeks, sensual moans, and loud supplications to the divine in the middle of a crowded New York deli. As Harry is frozen mid–pastrami sandwich in a state of mortified disbelief, an elderly woman at the next table nods to Sally and tells the waiter, "I'll have what she's having."[1]

Ryan's performance may have earned its place in Hollywood history, but it wasn't her bowl of Katz's coleslaw that sent her into paroxysms of pleasure. If her orgasm was indeed startlingly realistic, then what kind of trick or technique, the viewer may ask, could send a woman into such a state of rapture? Modern media would have you believe that the road to the female orgasm is as challenging and treacherous as the path to Mount Everest's summit, and that those who seek to reach the peaks of passion must be as nimble as the most seasoned Sherpa. *Find your clitoris! Buy this sex toy! Relax! Get excited! Practice alone! Do it with your partner!* The commands of magazines, websites, and social media are often contradictory, sometimes bewildering, and—with the advent of increasingly sophisticated toys that promise pleasure with every buzz, suck, and twirl—increasingly expensive. But have we been missing the point? Is it possible to reach the Big O simply by stimulating the Big U—the uterus?

Some early sexologists would have said yes. Husband-and-wife team Irving and Josephine Singer wrote a number of texts in the 1970s analyzing and classifying what they considered to be the three types of female orgasm: the vulval orgasm, achieved by either clitoral stimulation or penetrative thrusting; the uterine orgasm, initiated by thrusting of the penis against the cervix (the lower part, or "neck," of the womb); or the "blended" orgasm, which had characteristics of each of the former types.[2] A later critique of this work points out that Irving Singer was "a philosopher with no experience of laboratory studies, but who analyzed the descriptions of orgasms from a limited literature."[3]

In fact, Singer's fascination with sexuality seems to have waned with time, or at least, it was selectively ignored by the academic community; at the time of his death in 2015, an obituary posted online by the Massachusetts Institute of Technology, where he was professor emeritus of philosophy, describes him as "focusing on topics such as the philosophy of love, the nature of creativity, moral issues, aesthetics, and philosophy in literature, music and film."[4] Like a proud parent glossing over their teenager's "awkward" phase in favor of celebrating their more socially acceptable achievements, the subsequent list of Singer's esteemed works omits all mention of his treatises on human sexuality. As for Josephine Singer, her 2014 obituary makes only oblique reference to the "many books" on which she collaborated with her husband, with just one of her works mentioned by name: "Fanny and the Beatitudes," an article she wrote for the Jane Austen Society of North America.[5] With Irving moving swiftly on from his relatively unsubstantiated studies of sexuality and his wife's interests appearing to have taken a more sedate turn, one could be forgiven for taking their "discovery" of the uterine orgasm with a large pinch of salt.

Some recent sources still point toward a grain of truth among that salty pinch. Regardless of the scant evidence for the uterine orgasm, the phenomenon still has its enthusiasts. One online article with the breathlessly optimistic title "There Are 8 Kinds of Female Orgasms— Here's How to Have Them All!" picks up where the Singers left off, running—nay, skipping, cartwheeling, and backflipping—with the notion of the uterine climax. "A cervical-uterine orgasm will feel deeper, more intense and yet more 'round' than the g-spot orgasm," explains the author, "and will be accompanied by strong emotions, love, oneness with self, partner and god, ecstasy and transcendence, tears, crying and a feeling of deep satisfaction on all levels."[6] After such a glowing endorsement, any reader could be forgiven for wanting to have—to paraphrase that deadpan bystander in *When Harry Met Sally*—what the article's author is having. However, with such a clickbait-y title, no scholarly source list, and the not-insignificant detail that the author in question is a man, it's possible that this promise of a euphoric cervico-uterine orgasm is too good to be true.

Perhaps a more reputable source might pour cold water on the theory of the uterine orgasm? Perhaps not. *Our Bodies, Ourselves*— originally a bestselling book penned by second-wave feminists, and now a mainly online resource—has been a go-to source of intimate information for people born into female bodies since the Singers' heyday in the seventies. The *Our Bodies, Ourselves* website suggests, "Some women find the cervix and uterus crucial to orgasm," while stopping short of any claim that the climax is caused by stimulation of the cervix itself. "An orgasm that some women describe as feeling 'deep' or 'uterine' is brought on by penetration of the vagina," writes the author.[7] Again, as with the Singers' work, no clinical evidence is presented for these claims. Indeed, there is scant proof—if any at all—that the cervix is a source of sexual pleasure. A rare study of this concept by a team from Rutgers University in 2012 found that cervical self-stimulation with a "rounded tip cylinder" appeared to activate sensory responses in the brain; a not entirely surprising finding, the authors note, given that the cervix is infiltrated by the hypogastric and vagus nerves. However, the authors also admit that their results may have been skewed by a kind of orgasmic cross-contamination: participants' pleasurable responses may not have derived exclusively from the cervix, but from indirect stimulation of the clitoris and/or vagina with the "rounded tip cylinder."[8]

Proof of the cervical orgasm, then, remains elusive, but perhaps the question itself is moot. Perhaps the emphasis on penetrative sex as an integral precursor to orgasm—be it vaginal or cervico-uterine—may be more of a reflection of dominant cultural narratives around pleasure than a factual representation of female physiology and desire.

-

Whether or not you believe that climax can be achieved by stimulation of the cervix alone, there is one incontrovertible fact about the womb and its role in orgasm. Numerous sources concur that the uterus—along with the vagina and the muscles of the pelvic floor—contracts rhythmically during orgasm, regardless of the source of stimulation. While most people will be unable to feel these uterine tightenings, any pregnant woman who's noticed her belly going hard and taut after orgasm can testify to the truth of the phenomenon. When your womb has expanded to accommodate a fetus, a placenta, and nearly a liter of amniotic fluid—and takes up almost all the space in your abdomen—those previously unnoticeable postcoital contractions are pretty hard to miss.

What, though, is the function of these orgasmic spasms? If we are to believe that our physiology evolved over hundreds of thousands of years with the sole purpose of perpetuating the species, then it stands to reason that the uterus's response to pleasure fulfills some important reproductive purpose. In her book *Bonk: The Curious Coupling of Sex and Science*, Mary Roach describes scientists' weird, wacky, and often downright zoological attempts to demonstrate that the womb's orgasmic action somehow pulls sperm deeper into the reproductive tract, thus increasing the chance of successful fertilization.[9] Roach's "Upsuck Chronicles"—as she calls this unusual chapter in scientific history—begin as far back as 1840, when a German anatomist observed the quick transit of sperm from a dog's vagina to her uterus, leading him to speculate on the ingestive powers of the canine climax. Nearly one hundred years later, scientists still had a thing about animal sex and sperm: in 1939, a team of researchers in Illinois stimulated rabbits to orgasm, "inseminated" them with dye, and then used fluoroscopy to show the subsequent swift flow of dye from vagina to womb.

Nearly one hundred years and many experiments with bunnies, bulls, mutts, and monkeys later, the "upsuck" theory has yet to be conclusively proven, although one study by German researchers in 1998 has perhaps come closest. In a painstakingly created simulation of human coitus, scientists injected "radiolabeled microspheres" into volunteers' vaginas before administering an intravenous dose of synthetic oxytocin—the "love" hormone that's released in orgasm (and labor), causing that holy grail combo of rhythmic contractions and powerful sensations. The researchers' technique seems to have done the trick, initiating quasi-orgasmic contractions. Using fluoroscopy to track the luminous faux-sperm on their journey, the scientists found some pretty impressive results: after the oxytocin was given, there was a "reversal of the pressure gradient from a fundo-cervical to a cervico-fundal direction." In other words, after being given a synthetically induced orgasm of sorts, the wombs began to "suck," subtly but effectively pulling sperm inward toward any potential egg that might be lying in wait. In conclusion, the study's authors write, "These data support the view that the uterus and Fallopian tubes represent a functional unit that is acting as a peristaltic pump."[10] For the first time, scientists appeared to have proven conclusively that the components of the female reproductive system work in fine coordination to facilitate the sperm's safe journey to the egg. The uterus and its tubes aren't just a vessel in waiting: they're actually taking an active role in the very earliest moments around conception.

More recent research has used cutting-edge technology to confirm the importance of uterine peristalsis—tiny, wavelike motions of the smooth muscle in the womb. In 2017, Dr. Belén Moliner, a reproductive medicine specialist at the Instituto Bernabeu in Spain, pioneered the use of 4D ultrasound imagery to observe what she calls "the amazing movements of the womb." These movements, she explains, are continuous throughout the uterus's monthly cycle, enabling the womb to shed its lining and then encouraging better sperm transport around the time of ovulation.

"Uterine peristalsis changes through the menstrual cycle, fulfilling its purpose," she says. "Firstly, it has high intensity and low frequency to release endometrium during the menstrual cycle . . . Then, the directionality changes and the peristalsis increases its frequency, which helps the transport of the sperm to the Fallopian tubes." As progesterone rises in the latter part of the cycle, the "peristaltic pump" is subdued; Belén explains that this hormonal effect is crucial in "helping the embryo not to be rejected and assisting it to implant during the secretory phase." Her ongoing work appears to confirm the relationship between healthy progesterone levels, effective peristalsis, and successful conception, opening up new possibilities for diagnosing and treating infertility.[11] The frontier of this field appears infinite, with an ever-expanding horizon: "Every day I discover new uterine movements that I don't completely understand. We analyze every video in detail, trying to gather all the possible information and figuring out what it means." While your eyes scan this page, Belén may be scanning her screen, searching for hidden patterns and clues among the cells of this precious pump.[12]

As with almost every advance in the study of the human body, every step forward seems to push that tantalizing horizon two steps farther away. As the uterus unravels its secrets—the currents that carry sperm on an ever-changing tide—the mystery of reproduction appears more inscrutable, more complex, more awesome than ever.

–

UTERINE UPSUCKS AND PERISTALTIC PUMPS: THESE CONcepts still seem outlandish, in spite of the growing body of evidence to support them. It's no wonder that we find the womb and its tiny waves so hard to understand: this notion that the female body plays an active role in fertilization turns traditional narratives about the process on their head. Emily Martin argues that "the picture of egg and sperm drawn in popular as well as scientific accounts of reproductive biology relies on stereotypes central to our cultural definitions of male and female." She explains how, from the earliest phases of sperm development to the so-called "race" to the egg and the triumphant moment of conception itself, the prevailing idea of male gametes as active heroes and their female counterparts as passive maidens "keeps alive some of the hoariest old stereotypes about weak damsels in distress and their strong male rescuers."[13] In describing sperm as "streamlined," "strong," and moving "with velocity," while the egg "drifts" aimlessly along the uterine tube until it fulfills its mission by being "penetrated," textbooks, clinical reviews, and popular literature superimpose a patriarchal framework on a physiological process that is more complex than any simplistic gender stereotype. As for the womb, were you ever taught about its role in the moment of conception? Me neither. In the public imagination, the uterus is merely a backdrop for this dramatic process: a stage on which the sperm struts his stuff.

If you've read this far, you might be able to guess what comes next: a revelation that the uterus actually plays a starring role in those first vital moments of baby-making. Reader, you are correct, but in order to understand how this happens, we need to go on a journey that starts in a town in Poland, reaches its thrilling climax in a hysterectomy clinic in Israel, and ends on a cliffhanger featuring monkeys, bats, and a disgruntled anthropologist in Switzerland. All roads, of course, lead to the womb.

—

HISTORY HAS NOT RECORDED THE NAMES OF THE twenty-five women who gathered in Vaclav Insler's lab in Tel Aviv in 1979. We don't know whether they were young or old, rich or poor; whether they were bronzed and sandaled *kibbutzniks*, arriving on buses from one of the many communal farms that thrived in the wake of the postwar Jewish diaspora; whether they were local women, taking a day out of their sophisticated urban lifestyle in Israel's most secular city; or whether they were pious *bubbes* in headscarves and long skirts, making a rare foray out of Jerusalem's ancient limestone walls. What we do know is that they all had one thing in common: they were booked in for hysterectomies. Not an unusual procedure—in fact, one of the most commonly performed operations in the world—but more unusual in the fact that each woman had agreed to be inseminated with a stranger's sperm just a few hours prior to the removal of her uterus.

What kind of women would agree to participate in such a strange experiment? Were they promised a hallowed place in the historical register of great scientific achievements, or was their consent gained briskly and briefly, with the insemination only mentioned in passing, just another pre-op add-on, as unremarkable as the requisite pubic shave and overnight fast? Were these women motivated by a desire to contribute to the greater good of their sex, or were they ticking a box? We'll never know the names, ages, occupations, or motivations of these women, although we do know now how their participation in such an unusual trial led to one of the most striking and underappreciated discoveries in reproductive science. We also know exactly what kind of man would dream up such a thing.

Vaclav Insler is described in a retrospective of his field as "a fine physician and human."[14] Like many Jewish scientists of his generation, Insler's achievements followed years of hardship: he was born in 1929 in Stanisławów, a thriving town (then Polish, now Ukrainian) that became the site of one of the deadliest Nazi massacres of the Second World War. Young Insler's escape from these horrors took him to Kraków, then to Hungary, then back to postwar Kraków, where he lived until he emigrated to Israel in 1957. In spite—or perhaps because—of his years spent evading death and disaster, Insler's subsequent career at Tel Aviv University saw him focus on the creation of new life. The study he published in 1980 was risky, imaginative, and illuminating— its findings as surprising as its unconventional methodology.

By removing and studying the participants' uteruses shortly after flooding them with sperm, Insler was able to demonstrate that the uterus itself—or, more accurately, the cervix, that lower portion or "neck" of the womb—appears to store sperm in over 20,000 tiny "crypts." Even more remarkable, Insler wrote, was that "the quality of semen appeared to be of critical importance to sperm storage. The percentage of colonized crypts and sperm density were severely reduced in patients inseminated with abnormal sperm."[15] Whether the cervix actually selected the best sperm or whether only the finest sperm could survive the storage process was unclear, but the significance of Insler's fundamental discovery was unmistakable. With the help of his twenty-five volunteers, Insler had observed a hitherto unknown but reproductively crucial process: far from being a passive vessel during conception, the cervix stores and then slowly releases the best, most viable sperm into the main body of the uterus—in some cases, over a week after insemination has occurred. This discovery cast doubt over two of the most centrally held beliefs about fertility: that conception was a "race" dependent solely on the vigor and stamina of sperm, and that this race occurred in a "fertile window" of only a few midcycle days on either side of ovulation. Vaclav Insler and his twenty-five anonymous volunteers discovered something that should have sent shockwaves across the Mediterranean and beyond, rippling through the scientific community worldwide. Instead, the discovery was more of a gentle drop in the ocean.

In the decades since Insler's revelation, science has made little mention of cervical crypts and their potential role in human reproduction. An internet search for recent work on the topic yields a brief YouTube clip from a 2008 Discovery Channel documentary on life before birth; for eighteen seconds, you, too, can watch a CGI sperm thrashing helplessly around what's meant to be a maze of cervical crypts, but looks more like the rubbery tines of a Tangle Teezer hairbrush.[16] There is no clear line of scientific inquiry picking up where

Insler left off, no screeds of studies elucidating and elaborating on his discoveries, no new understanding of the cervix's role in widening the fertile window and its implications for couples struggling to achieve (or avoid) conception.

Whether Insler was disappointed by this turn of events remains a mystery. He died in 2013 after a long and varied career in maternal-fetal medicine; cherished, his biographer writes, by "thousands of women who remember him dearly and whom he helped tirelessly."[17] One man, though, is still angry about science's failure to recognize and build on Insler's discoveries. Forty-one years after those fateful hysterectomies, and nearly two thousand miles away from Tel Aviv, Bob Martin is angry. He's a big fan of Insler and an even bigger fan of the uterus, and from his book-lined home office in Zurich, he tells me why he thinks science is missing a trick.

"Nobody's ever taken it seriously when it comes to humans, and yet it has huge implications," Bob tells me, his voice edged with incredulity and frustration. I'm video-calling him after reading a blog in which he argues that almost everything we think we know about fertilization is a "macho myth"—a premise that's tantalizingly unorthodox. Bob—or Professor Robert Martin, emeritus curator of the Field Museum of Natural History in Chicago and adjunct professor at the University of Chicago and Northwestern University, to give him his due—has spent nearly sixty years in the field of biological anthropology, studying tree shrews, lemurs, chimpanzees, and humans, and virtually every mammal in between. His life has been one long quest for knowledge about what sets our species apart, and what we can learn from our furry cousins. The one thing that's blindingly obvious, Bob tells me as a pair of monkeys in the photo over his shoulder fix me with inquisitive stares, is that females of many mammalian species can store sperm in the cervix or in the main body of the womb for optimal fertilization a few days or even months down the line. "There are lots of well-established cases [of sperm storage] in other mammals," he explains. "It's routine in bats. There's nothing technically difficult about it. Some bats mate in the autumn and give birth in the spring, but that's too long for their short gestation periods, so the bats store the sperm, and fertilization takes place four months after copulation."

It's alarming to think about the implications this kind of physiology would have on human relationships. Consider, if you will, the shock of a woman who finds two lines on her pregnancy test on a rainy day in April, four months after a drunken clinch at the office Christmas party. Imagine, too, the challenge of accurately establishing paternity of a child who could have been conceived at any time during a summer of whirlwind romances.

"Sperm can be stored for four months [in bats] with no problem at all," Bob continues. "And you know, there are even some nice stories from insects about sperm storage. It's out there."

Many people might prefer not to think about the copulatory habits of creepy-crawlies—even the "nice" ones—but Bob suggests that our closest and cuddliest primate cousins could provide evidence of sperm storage, too. "There's good evidence from the rhesus macaque that there are crypts in the cervix and in the body of the womb. Any time during the last forty years, somebody somewhere with a macaque colony could have investigated this. Why didn't they? They investigated everything else. Why didn't they look at this?"

Reader, if you are that person who's gallivanting through life with a spare colony of macaques and you've failed to investigate this groundbreaking phenomenon in primate reproduction, shame on you. Professor Martin thinks this is a willful and potentially dangerous oversight of the active role of female bodies in the reproductive process. This failure to expand on Insler's work may be connected, he argues, "with this embedded philosophy that the male is the active part and the female is the passive part in this. Sperm storage is not something the male is responsible for, and the overemphasis on the male could have been a major factor in people not bothering to continue with this [area of research]."

The scientific establishment's oversight is symptomatic of a bigger problem: an unrealistically simplistic view of the female reproductive system. It's far easier, Bob says, to imagine women's bodies as identikit units, all ovulating, conceiving, and gestating in exactly the same way, than it is to develop a model that incorporates individual, variable processes like sperm storage and release.

"Men have seen everything as beautifully predictable," Bob explains. "I call it the egg-timer model of the female cycle because people have regarded it as almost like clockwork. You see all these graphs [of the menstrual cycle] showing a rise in estrogen, and then you get progesterone rising after that. If you take fifty women and average their cycles, that's what you get. But if you look at each individual woman, you get a certain pattern that is very different. These averages are a very mechanistic approach."

Of course, science is full of researchers barking at each other about glaring omissions and differences of opinion; I spent years at the family dinner table listening to my own father, a biochemist, recount this colleague's petty slight or that bench-mate's personal vendetta, often motivated by competition for scarce funding or academic prestige. As Bob continues his diatribe against what he sees as a major blind spot in the field of human reproduction, I soon realize that this campaign is not a garden-variety grudge. His frustration is rooted in a sense of compassion for the many thousands of women and their partners for whom conception is a struggle. As those black-and-white monkeys continue to gaze balefully at me from his office wall, Bob seems equally pained by the suggestion that a better understanding of the uterus's role in conception could have spared many women the agony of repeated cycles of IVF. In spite of over forty years passing since the birth of the first "test tube" baby, the process still carries an average success rate of only 20 to 30 percent, with many couples embarking on numerous cycles at great physical, emotional, and financial expense.

"The thing with IVF and with artificial insemination—in fact, with any assisted reproduction—is that it hasn't really gotten any better. I mean, people are refining the techniques and so forth," Bob says. "They've made minor advances. But we're still in a position where three quarters of the time, it doesn't work. And that tells me that we don't really understand what's going on with sperm storage and things like this. The fact of the matter is, it's a challenging process. It's awful for women to go through IVF. And to only get a twenty-five percent success rate is not very encouraging. And that's where we should be devoting research to make that better. You know, why is it only twenty-five percent? You would think just putting some sperm together with an egg shouldn't be that problematic, right?"

Wrong. It may be convenient to ignore the messy variety of women's bodies, and an outdated patriarchal narrative may persist in promoting the sperm as a dynamic "vital spark" and the female genital tract as a passive vessel, but these oversights do women a grave injustice. The science is there, telling us that the uterus—from its neck to its body, and all points in between—has an important role to play in the creation of life, if only we are brave and generous enough to listen . . . and if only we've got a macaque or two to prove it.

"So," says the doctor with the classic intake of breath that precedes a devastating diagnosis, "we have a few things to discuss here." It is season 7, episode 4 of the hugely successful hospital drama *Grey's Anatomy*, and main characters Meredith and Derek are waiting with bated breath to hear the reason for their recent miscarriage. Cut to a furious Meredith pacing the street outside the clinic a few minutes later.

"Hostile?!" she rages. "Did she really call my uterus hostile?"

"You're just getting stuck on the word," counters Derek, pointing out that the doctor also offered a plan of treatment, but Meredith will not be placated.

"How would you feel if she called your penis angry, or snide?" she rails.

Ever the optimist, Derek playfully points out how much fun it will be to keep trying for a baby.[18]

And just like that, Hollywood achieves a hat trick of medical misinformation: (1) that a womb can be hostile, (2) that "hostile uterus" is a valid clinical diagnosis with an effective, recognized course of treatment, and (3) that women are irrationally focused on those trifling little things known as "facts" instead of embracing a much more socially acceptable obsession with "sex."

Reader, we could have a fantastic discussion of the uterus's role in sex and conception by focusing only on the miraculous minutiae of myometrial contractions and cervical crypts. We could glory in scientific discoveries of the former, and we could conclude with a satisfyingly righteous rant about the latter. But the fact of the matter is that many women have little knowledge of the weird and wonderful actions of their womb in the very first moments of life's creation. Many women think—usually because they have been told so by a medical professional—that their uterus is inherently flawed, structurally hopeless, and, in fact, downright angry at any potential sperm that might dare to traverse its briny depths. Like Meredith, these women have been tarred with one giant, unhelpful, misogynistic brush, a kind of catchall, lazy substitute for genuine investigations or explanations. These women are told that they have simply drawn the short straw of gynecological evolution; they've got a lemon instead of a peach; a damp squib instead of a hotbed of fecundity. They have, in common pseudo-medical parlance, a hostile uterus.

But if a doctor says it, you could be forgiven for thinking, it must be true! Indeed, even the most cursory browse of online science blogs, fertility websites, and forums for those "TTC" (trying to conceive) would have you believe that a so-called hostile uterus is one of the foremost factors in female infertility. While it's true that certain physiological issues—scarring from fibroids or inflammatory disease, for example, or cervical mucus with less than optimum consistency or acidity—can make conception difficult or impossible, each of these is an individual clinical diagnosis in and of itself, with its own unique course of management. "Hostile uterus," on the other hand, is not a widely accepted condition or disease, and a literature search of reputable journals reveals only a few references to the term, mainly in the context of animal studies.

Like Meredith in *Grey's Anatomy*, people who have been told they have a hostile uterus may initially be most struck by the term's emotively negative connotations. After all, a uterus is an organ, isn't it, just like a heart or a kidney, and how can an organ have feelings or intentions? Is the uterus capable of some kind of capricious malice, snarling at every penile intruder? Journalist Caitlyn Goerner describes her confusion in a real-life situation similar to that of Meredith and Derek: "The one thing I hadn't prepared for—something I had never in my life heard of—was the news I actually received. 'You have a hostile uterus.' My blank stare was followed by a maniacal giggle . . . It's just that I can't name another body part with personified qualities. I imagined my uterus with a knife or a finger on a detonator. My uterus is a terrorist, you guys."[19] Goerner goes on to explain how it was actually her cervical mucus, and not her uterus, that was, apparently, the problem. By that point, though, the damage was done—the "hostile uterus" bomb had been detonated, with all of the blaming, shaming shrapnel it often brings.

In the course of researching this book, I spoke to other women who'd been tarred with the "hostile uterus" brush for a variety of reasons, but each had the same initial response of bemusement and self-doubt. One respondent on social media said, "After I had a termination of pregnancy, I was told by the doctor that my uterus was hostile because it was retroverted, but when I went away and did my research, it seems it isn't hostile at all! There's no major difference, really. Even though I wasn't having that baby, it felt a bit off to be told a part of me had been 'hostile' toward the pregnancy." In that instance, not only was "hostile uterus" used inaccurately—with a retroverted uterus very rarely having any impact on successful pregnancy—it also appears to have been used as a kind of judgmental stick with which to beat a woman already made vulnerable by the nature of her circumstances.

There are, of course, instances in which some abnormality of the uterus is a factor in infertility. But if and when this is the case, perhaps "hostile" should be replaced with a less emotively evocative word, or perhaps we can do what we seem to be capable of doing in every other area of medicine: we can use specific, accurate, appropriate terminology to explain the situation. After all, as Meredith points out to poor old Derek, when has a penis been called angry or snide, or otherwise anthropomorphized by a medical professional?

A plea, then, for a narrative around fertility and conception that's emotionally neutral and clinically accurate. We have seen how the womb responds to orgasm, and how it draws in, stores, and releases sperm, possibly even directing it toward the most favorable location for finding an egg. When that doesn't go to plan, blaming it all on a hostile uterus causes needless distress, denies patients the dignity of an accurate diagnosis and treatment, and is bound to make an already vulnerable person more confused, upset, and yes—even hostile.

Pregnancy

PLACENTAS AND HEARTBREAK PREVENTION

look down!
that space was a house!
heated perfectly with blood and food

—Hollie McNish, "belly"

Time, then, to approach the pregnant uterus, or "gravid" uterus, to give it its medical name. Gravid: the word sounds serious and weighty. Loaded. It is right to pause and take a breath before we lay eyes, hands, and minds on this organ in its glory, this poised and powerful muscle, this cradle, truly, of civilization. In my ten years of midwifery, I've made this approach countless times, and always, always, in that moment between drawing back the blue cubicle curtain and introducing myself and my intentions, there is a meditative pause. Perhaps, if the woman is on all fours with the sliver of a newborn head glistening between her thighs, the pause lasts the merest fraction of a second. No matter how big her belly is: it is doing its thing. If the woman is resting comfortably, or telling her boyfriend on the phone how to navigate the hospital corridors, or sleeping, or eating a bucket of fried chicken—the circumstances as diverse as the women who arrive in my care—there is less urgency, and the moment may allow for contemplation.

Even the most novice student midwife knows that inspection—a visual assessment of size, shape, color, and motion—precedes palpation. Feel your way with your eyes before you touch with your hands. Is that bucket of chicken resting on a neat basketball bump, or can you see a belly so swollen and vast that it bursts through an unbuttoned fly and a too-small T-shirt, the woman's head only just visible behind this massive globe of flesh? As she pulls up her shirt and you draw closer to the bedside, are you brave and kind enough to read the stories etched on her skin? A first-time mother's red tiger-stripes of freshly stretched skin, or the silver striae of a "multip"—mother of many. The faded smile of an old Cesarean section scar; a dolphin tattoo arcing gracefully over the bony crest of a hip; a jeweled hoop winking from a belly button; bullet wounds sustained in a far-off war. Before we have even touched this woman, we know a little of her life: the triumphs and tragedies through which she has carried her womb, a cargo silent but vital.

Now, then, you are close enough to smell her breath: mints, or masala, or the pear-drop fug of ketones. Your legs press against the side of the bed as you ask if you can touch her—consent, always, first and last, more vital to this contact than any instrument or advice—and finally, the pads of your fingertips rest against her abdomen. Lightly, now, so that you can almost feel each whorl of your thumbs making contact with her skin, then firmly, feeling for the tone and shape of the uterus, and the parts of the passenger within it. Before twelve weeks of pregnancy, the womb is still small and secret, tucked beneath the pelvis as it does its early work, but then at eighteen weeks, your hand glides down and there it is, the size and shape of the sleek, round head of a cat, nestled pleasingly under the curve of your palm.

Or perhaps this pregnancy is more advanced, the uterus expanded to its full size, filling the woman's torso from the notch between her ribs to the hard arch of her pubic bone. With one hand grasping the soft, broad bump of the baby's bottom, your other hand glides down the abdomen, coursing along the firm ridge of the fetal spine until it meets the familiar bulk of a head swinging gently at the brim of the mother's pelvis like a pendulum, marking time. Perhaps, suspecting an excess of amniotic fluid, you tap one side of the woman's belly and notice the characteristic rippling "thrill" confirming your suspicions. Perhaps her abdomen swells and relaxes with early contractions, or perhaps your heart skips tightly into your throat as you realize this uterus feels rigid and wooden: the unmistakable sign of a concealed bleed. Somewhere under your hand, the placenta is shearing quietly away from the wall of the womb. Your fingertips have told you how this story starts; your brain, whirring, alive, trained for every eventuality and emergency, knows how it ends. In that moment, you and the uterus and the person who carries it are one: gravid, serious, weighty.

But again, we must pause, rewind, return to that first intake of breath. Like all keen students, we are getting ahead of ourselves. We will come to the drama of labor—the heat and the noise—and we will peer through that cosmic rupture in the space-time continuum when one person becomes two. First, though, we must begin at the beginning: return to that very first flicker of life, when egg and sperm merge into a blossoming cluster of cells. Now, unseen and unfelt, long before two lines appear on a stick and a knot of nausea takes up residence in a mother's gut, the uterus performs some of its mightiest magic. Like a trapeze artist swinging through the air and grasping with blind faith for the hands of her partner, the newly fertilized egg reaches out for the blood-rich bed of the womb, and the womb, in turn, gives everything it's got. Blood, oxygen, nutrients, immunity; life.

IN A SLOW-MOTION JOURNEY REPEATED BY UNTOLD GEN-
erations, the zygote—as we must call this new union of sperm and
egg—drifts through inner space until it finds a foothold in the lining
of the uterus. Within these first few limbo-like days after conception, the zygote takes another form, and another name: a blastocyst,
whose inner layer will form what we'll eventually recognize as a
baby, and whose external cells will transform into the placenta and
chorion (the outer part of the "bag of waters" in which the fetus develops). What happens next is one of nature's wonders: an exquisitely
tuned exchange of biochemical and immunological signals in which
the uterus allows itself to be invaded by the blastocyst, becoming
first an open wound, then forming an elaborate nutritive network
of intricately intertwined glands and arteries, and finally creating a
completely new organ within an organ—the placenta—one which
will nourish and sustain the fetus until its first gasping moments of
contact with the outside world.

Common sense tells us that any misfires or malfunctions at this
stage can prevent implantation, causing what could be referred to as
a missed period or an early miscarriage, depending on one's definition of pregnancy, or even of life itself. What has only become apparent in more recent years is the effect of flawed implantation on
complications in the later stages of gestation. If only scientists could
create a kind of synthetic mini-womb, the crucial processes of implantation and placentation could be studied with ease. We could observe at close quarters the myriad cells and substances that allow the
endometrium—the lining of the womb—to welcome the burrowing
blastocyst: the immune-regulating B-cells and T-cells, and even the
ominously named natural killer cells that allow just the right amount
of localized inflammatory response, no more, no less; the nutrient-
rich uterine milk secreted by hardworking glands; the spiral arteries
that flood the endometrium with blood and the tiny, pooling lakes

that absorb the waste excreted by the blastocyst in return. If only we could magic up a sci-fi endometrium—a perfect replica of uterine tissue in miniature, a model that would behave in the lab just like the real thing behaves in life—then we might finally illuminate some of early pregnancy's darkest corners.

-

THE UNIVERSITY OF CAMBRIDGE'S DEPARTMENT OF PAthology is an imposing redbrick building, with its pillared and crested facade. If we travel up the staircase, gliding past polished wooden banisters and portraits of pathologists past, we find ourselves in a lab—gazing across long benches with high-powered microscopes, sheaves of paperwork, and humming fridges stickered with biohazard warnings—where Dr. Margherita Turco and her colleagues are making gynecological history.

"I became really interested in how different cell types are differentiated in the sheep embryo," Margherita tells me. We are sitting in her office next to the lab, and the buttery sunlight of a midsummer morning streams through a window behind Margherita, illuminating her with a fuzzy halo as we speak. She is the most modest, softly spoken figure I have met on my uterine travels. I worry that my recorder won't pick up the quiet tones of her voice; I fiddle with the volume button while Margherita tells me how it all began with sheep.

"I actually studied veterinary biotechnology because I was interested in using reproductive technology to help endangered animals," she says, "but it's really hard to get into that field and to find money to do that. Then my partner came across a job advert for a postdoc [post-PhD research position] here to study the stem cells of the placenta and to develop a model."

The model in question wasn't just some kind of Meccano mockup, nor was it the standard slice of plated tissue that scientists have been studying for centuries. What Margherita aimed to do—and what she has now done with success beyond her wildest expectations—was to create a placental "organoid," a tiny 3D version of placental tissue that, under the right conditions, could be infinitely reproduced, treated with any number of hormones and drugs, and analyzed for its responses and behaviors.

"A group in Holland had established this model system, called organoids, of the intestine," Margherita says, "and it revolutionized how we can study human tissues. Instead of just plating cells as a monolayer in 2D, you try to re-create the environment of the tissue by creating a 3D environment for the cells to grow. You isolate cells from a normal or diseased biopsy, you put them in this 3D droplet, and then if you give them the right signals, they kind of do everything themselves. They self-organize, they assemble, they organize into these miniature structures."

While researchers around the world had pressed ahead with developing organoids of many other bodily systems, Margherita soon realized that—in spite of placentation being arguably one of the most important physiological processes of the human body—finding money and support for the development of gynecological organoids was going to be an uphill battle.

"When I came into this field, it was really clear that it was hard to find funding," she says. "It's not like getting money to study cancer, heart disease, or the brain. The placenta? It's mostly like, 'Who cares? We just throw it away.'"

At this, I think guiltily of all the tepid, just-birthed placentas I have bagged and thrown into the labor ward sluice, disposing of precious tissue that could have been alchemized into scientific gold.

"I was very lucky that I came to work here where they put money into this kind of high-risk product that no one else would ever fund."

As with so much in life, the endeavor with the highest risk often yields the greatest reward, and Margherita's mission was no different; she aimed to identify and understand what researchers refer to as the "maternal–placental dialogue," that incredibly complex but vitally important exchange between the motherly host and her startlingly invasive placental parasite.

"It's a dialogue," Margherita explains, "because on one side, the endometrium secretes and somehow stimulates the placenta, and that, in turn, allows the placenta to send more signals to the endometrium. So they're kind of mutually signaling to each other, to establish this relationship, but we have no idea what that is." You read that correctly: no idea. While money has been plowed into solving medicine's more obviously pressing (and financially sexier) problems, the very act by which a person's body allows the creation and nourishment of new life has been largely overlooked and underfunded.

The maternal–placental dialogue has fallen on deaf ears, and to devastating effect. Any glitches in this early exchange of messages can contribute to or cause a set of problems that clinicians often call—with reverence that seems somewhat poorly judged, given the circumstances—the "Great Obstetrical Syndromes." This catalog of catastrophe includes preterm birth, preeclampsia, miscarriage, and stillbirth: conditions and events which affect millions of families of every country, creed, and color each year. Ultrasound scans, blood tests, and invasive investigations can sometimes detect life-threatening problems and treat them before they reach their tragic conclusions, but all too often, families and clinicians are left to search for clues after their worst nightmare has already become reality. In many cases, the causes of these tragedies can only be determined in retrospect: pathologists can examine the placenta and may be able to point to what's gone wrong, or the body of a stillborn child may offer up its truths at postmortem. We listen for the echoes of whispered clues long after that crucial early dialogue between mother and child has ended. In many cases, this process poses more questions than it answers, and the not-knowing is almost as painful as the event itself.

Margherita and her colleagues theorized that if we could only build and study organoids of the placenta, we could watch how these 3D models succeed or fail in proliferation and implantation, and we might potentially stop some of those "Greats" in their tracks. With what I suspect is characteristic understatement, Margherita glances down at her screen and admits that producing a placental organoid was "really, really challenging. There were just so many things we didn't know. Trying to create a model of a tissue that you don't understand very well is kind of like rummaging in the dark."

Over time, though, she developed a successful method—a recipe of sorts—for creating these novel models.[1] In a twist of bittersweet irony, the raw material came from first-trimester pregnancy loss tissue, donated by women at a nearby hospital. This tissue, composed of a mix of endometrial and early placental cells, was then treated with enzymes, and the placental cells embedded in a small droplet of Matrigel: a kind of gelatinous soup, full of proteins and growth factors that mimic the body's own environment.

"Each cell of each tissue requires different signals to grow," Margherita says, "and then once we identify the right signals—the right cultural conditions—after a few weeks, you start to see the formation of these little globular structures which essentially represent the villi [early branches] of the placenta. You can take this droplet and measure all the organoid's products; there are lots of really interesting products secreted into circulation that we still don't know anything about." Once again, you read that correctly: we still don't know *anything* about many of the mysterious substances released by the very early placenta. In the twenty-first century, with all the sophisticated science of the modern world, so many of these crucial signals—this biochemical love language that heralds either new life or early loss—

remain indecipherable. The creation of these organoids that could be reproduced and studied ad infinitum was, Margherita modestly admits, "really exciting. But as a scientist, you're always thinking, what's next?"

What happened next was even more thrilling than Margherita had expected: she discovered a surprising but equally valuable by-product of placental organoid creation.

"As I was developing the placenta model, I kept seeing this other cell type emerging in culture, and then some of its markers," she recalls. "It didn't quite look like trophoblast [invasive early placental cells]. It looked really different." These strange-looking outliers were, in fact, endometrial organoids: tiny 3D versions of uterine tissue that had developed in the nutrient-rich Matrigel along with their placental sisters.[2] Further analysis revealed that these endometrial organoids could be treated with progesterone and estrogen and would behave just like uterine cells would *in vivo*—in real life. From those samples of pregnancy loss tissue—usually sent to an incinerator as clinical waste, or passed into a toilet in moments of private sorrow—Margherita had grown little clones of both placental and uterine tissue, clones that could be scrutinized, treated with hormones and medications, replicated again and again, and used as a kind of obstetric Rosetta Stone: a key to that dialogue whose true meaning has evaded researchers for so long.

Margherita tells me that her dream now—and one which, she suggests, could become a reality in our lifetimes—is to create a mass biobank of endometrial and placental organoids derived from tissue donated by women of all ages, ethnicities, sizes, and disease states, and to use these organoids to develop treatments and interventions with the potential to improve fertility and prevent the Great Obstetrical Syndromes. A Dutch team have already succeeded in testing individualized cystic fibrosis therapies on organoids derived from patients' own rectal biopsies.[3] Margherita tells me that similarly bespoke IVF regimes could potentially be trialed on women's endometrial organoids in this way, finding each person's window of implantation and understanding their unique response to the kinds of hormones that are usually administered with a risky and expensive one-size-fits-all approach.

"Can we tailor the hormonal regime a little bit more to the woman instead of just giving one generic treatment and then crossing our fingers?" she asks. "It's promising," she says. "It's exciting."

Margherita's voice becomes brittle as she rails against the soaring costs of IVF treatment and the expensive and often unproven "add-ons" that are sometimes marketed at vulnerable women and their partners. "I find it so awful. I myself had to think of it because I'm expecting my second one now," she says, passing a protective hand over her abdomen. "I have my first boy, who's ten, but it took us so long [to conceive again]. And, you know, I was studying all this, and thinking, it was so easy the first time, why is it so difficult now? What happened? What changed in the endometrium between then and now? And then last year I had a miscarriage. I had already kind of given up. Finally, we started considering IVF. And when I started looking at clinics, even as a scientist, I was like, this is so scary. You know, what are these things that they're offering? What is the evidence? So I really feel like these should be informed things. You know, you pay for what you get, and if you can, you get the best treatment with less stress."

In another universe, I think, Margherita could have collected her own tissue, created her own placental and endometrial organoids, and tested them and treated them under the microscope at length until she finally understood what had gone wrong, and why. She could even have tweaked her perfect fertility regime, administering this much progesterone and that much estrogen, watching the cells proliferate and differentiate, measuring their secretions, and timing their changes until she understood exactly which part of the dialogue had been missing. She could have filled in the blanks; kept that cellular conversation going in the lab, and then replicated it in real life. Instead, and until this technology is better funded, resourced, understood, and available to us all, there is the Russian roulette of IVF, and all too often, there is heartbreak. There are the women who enter my hospital's doors in hope, and leave with empty arms and endless questions. Around the world, there are countless bathroom bins brimming with negative pregnancy tests; there are tears when those first spots of blood appear on clean cotton gussets; there are postmortems for perfect, fully grown babies, and there are midwives whose children wonder why they get hugged especially tight after certain shifts, with the meager explanation "just because" mumbled into their hair.

And yet, as Margherita says, her work is promising, and there is hope, and nowhere is that hope more evident than among the benches of the pathology lab. It may be stereotypical to suggest that a female-only environment is inherently nurturing, warm, and fuzzy (a notion I know to be false from my time in midwifery, a field that can be as macho and cutthroat as a trading floor or a building site), but as Margherita shows me around her lab, the goodwill is palpable. White-coated women move among the microscopes with calm, quiet purpose. A sticky note on the worktop announces "I LOVE YOU XXX" to some anonymous recipient; a researcher guides me gently toward her microscope, adjusting the focus so I can view a slide of tiny organoids suspended in their soupy gel; another casually mentions her own history of miscarriage—"It was no big deal"—before talking me through her current project. Realizing that it's almost time to meet my daughter, who is waiting for me a few streets away—my own maternal obligations never far from my professional ones—I make my excuses, but not before Margherita gathers the researchers around me for a group photo. Shoulder to shoulder with the team, our huddle feels like an embrace, even though the current wave of Covid forces us to keep a salutary distance between our bodies. One researcher apologizes for her "silly" pink mask, which I tell her my daughter would have loved; another tells me she's halfway through reading my last book, a memoir of my midwifery career thus far. I say that she can think of me next week, when I'll be reimmersed in the thick of my day job. And when I am back in the maternity hospital, uniformed and aproned, clothed in layers of polythene and worry, I remember them, too—those white-coated women at their benches, which I realize I've now begun to think of in my head as the Heartbreak Prevention Lab. The women in my care come to me heavily pregnant, their bellies taut with fat, wriggly babies and placentas as big as dinner plates, but it all starts with those tiny cells, whispering to each other, sending secret signals, exchanging rich crimson blood and uterine milk and all manner of mysterious messages. It's a dialogue we have yet to decipher, but Margherita and her colleagues are listening.

Tightenings

BRAXTON HICKS AND THE IRRITABLE WOMB

Then, through the infant's enforcing, and the paines of the mother, the womb openeth . . .

—PERCIVAL WILLOUGHBY (1596–1685), *OBSERVATIONS IN MIDWIFERY; OR, THE COUNTRY MIDWIFE'S OPUSCULUM OR VADE MECUM*

As remarkable as the uterus's inner workings may be in the early weeks and months of pregnancy, only one change is visible to the external observer: the expansion of a woman's abdomen as the womb inside her stretches and grows. A mother may notice that the jeans that fit one day barely button across her belly the next or, as happens so frequently, friends, colleagues, and even strangers may feel suddenly compelled to comment on her growing girth. "You've popped overnight!" perhaps, from another parent at the school gates, or "I'll have to charge extra for twins," with a nod and a wink from the bus driver who's heretofore been glassy-eyed and mute. Humanity's fascination with the transformation of pregnant bodies seems to be innate and perpetual, from primitive fertility figures with their pendulous breasts and bellies to modern-day tabloid spreads of celebrities cradling their "bumps" in carefully staged photo shoots. Simply by expanding, the womb transforms its owner's body from private to public, from sexual to maternal, inviting us as individuals and as a society to project our views and values on the mother as she changes before our eyes. By the time a pregnancy has reached its full term, the gravid uterus is almost large enough to obscure our view—literally and figuratively—of the woman herself.

Just as a person's identity is more multidimensional than that of the cliché "blooming mother," so the womb's expansion is more complex and sophisticated than the mere fact of a widening waistline. The myometrium—the main muscle layer of the uterus—is a richly textured basketweave of supportive and elastic fibers capable of growth, contraction, and relaxation depending on time and circumstance. Highly vascular and flooded by surging tides of estrogen and progesterone, myometrial cells can stretch up to fifteen times their original length, allowing for the growth of the tiny human inside it from microscopic dot to fully fledged infant. As the fetus gets bigger, so, too, does its uterine home: the womb becomes simultaneously thinner and heavier, a kind of magically distorted fun-house-mirror reflection of its former self. The myometrium can be less than 2 centimeters thick at full gestation (roughly forty weeks of pregnancy), while the bulk of the muscle itself can increase by as much as tenfold, from about 100 grams at conception to over 1 kilogram at the onset of labor.[1]

Labor: the obstetric endgame, the grand finale of gestation, the moment we've all been waiting for. We all know what the uterus does then, don't we? If all goes well, it contracts, dilates, bears down, and ejects a baby. There's that other bit that's often forgotten, too—the messy business of expelling the placenta and amniotic sac—achieved by more contractions in which the uterus cleverly clamps down on the vessels in its interior wall, leaving mother and midwife with just a trickle of blood and a satisfied smile. End of story. Or is it?

As with virtually every other aspect of the female reproductive system, the truth is a little more complicated. Labor isn't really the end, nor is it the beginning, and it wouldn't really be fair to call it the middle. What's more, contractions don't just happen in these final days and hours of pregnancy, and their presence doesn't always indicate that birth is imminent. In fact, most of the time, it isn't. Confused? Join the many shamans, wise women, midwives, and obstetricians whose minds have been boggled by the uterus's most famous yet unpredictable party trick.

IN A PHOTOGRAPH TAKEN IN 1873, JOHN BRAXTON HICKS cuts a dashing if slightly eccentric figure: a distinguished gentleman, and a medic in his prime.[2] Great, wisping mutton-chop whiskers fan out from his cheeks and join under his chin like a woolly strap. Gazing up at a distant point just above and beyond the photographer's hooded figure, perhaps the namesake of one of pregnancy's most infamous conditions is already envisioning his legacy. Maybe he knows that women everywhere will someday clutch their tightening bellies and curse his name (or almost his name, as is the case with the many women who phone the hospital where I work to report that "I'm having those Branston Hicks").

From his early childhood in Sussex to his later years of medical practice, Braxton Hicks was fascinated by the mysteries of the natural world. While advancing his career, first in general practice and then in obstetrics at some of Victorian London's most eminent hospitals, he continued to collect all manner of flora and fauna, from lichen to earthworms. Every living creature—small or large, beautifully simple or stunningly complex—captivated Braxton Hicks, who was described by his admiring peers as "an amiable man with a cheerful expression, and bright piercing eyes."[3] (The good doctor also amassed a substantial collection of Wedgwood china. His eye for tableware was not, perhaps, an indicator of keen obstetric insight, but let's take a moment to appreciate this facet of the man's eclectic personality.)

From his humble beginnings as an assistant obstetric physician at Guy's Hospital in 1858 to his appointment as that institution's consultant physician twenty-five years later, John Braxton Hicks became known as a prolific, inquisitive practitioner, keen to push the boundaries of what was still a relatively new role. Childbirth had been seen as an exclusively female domain of care until, in the seventeenth and eighteenth centuries, "man-midwives" or accoucheurs began to attend the births of French and English royals. Having a man at the foot of one's bed—potentially with a shiny new invention like forceps in his hand—became a sign of status and sophistication available only to the very wealthiest society doyennes. Thus, the term "obstetrics"—quite literally, the study of "standing in front"—was born. Traditional lay midwives were increasingly viewed by the moneyed elite as primitive and unclean, while men like John Braxton Hicks rapidly gained control of the birth space (and of the bodies that labored within it).

Braxton Hicks appears to have had a particularly meteoric rise through the medical stratosphere, publishing 133 papers during his career. Perhaps the most influential of these scholarly works relates to Braxton Hicks's observation that painless uterine contractions can occur throughout virtually every stage of pregnancy without actually causing cervical dilatation.[4] While one suspects that this characteristic of the womb had already been experienced and acknowledged by generations of women since time immemorial, Braxton Hicks was the first white, moneyed Western man to identify and investigate the phenomenon, so of course, like a father christening a favored firstborn, he gave it the best blessing he could: his own name. "Braxton Hicks contractions" is now a common term in pregnancy parlance; however, if women's lived experience had carried as much weight as male-dominated academia, these painless tightenings could easily have been named after any of the millions of mothers around the world who were already well aware of their existence.

As the first prominent man among men to explore this concept, Braxton Hicks's ideas met with resistance; his senior colleagues held the long-established opinion that the uterus only begins to contract of its own sudden, unpredictable accord at the onset of labor. Braxton Hicks had to remind his peers that labor could and often did occur at any point in pregnancy, proving the uterus's ability to contract at any time. In the words of the man himself:

> *It was a source of difficulty to the older obstetricians to explain how that, at a certain time, namely, at the full period of pregnancy, the uterus, passive up till then, began all at once to acquire a new power, that of contracting; forgetful that, long before the full period had arrived, the uterus has the power to expel the foetus, and under mental excitement or local stimulation, attempted to do so frequently. But after many years' constant observation, I have ascertained it to be a fact that the uterus possesses the power and habit of spontaneously contracting and relaxing from a very early period of pregnancy, as early, indeed, as it is possible to recognise the difference of consistence—that is, from about the third month . . . The constancy with which these contractions of the uterus have always occurred to me leaves no doubt on my mind but that it is a natural condition of pregnancy irrespective of external irritation.*[5]

These observations may have been fairly radical in their time, but Braxton Hicks's suspicions have now been borne out by years of research and close observation: we know conclusively that the pregnant uterus begins to have mild, irregular contractions from as early as six weeks of gestation. These events are often imperceptible until the second or even third trimester of pregnancy; by this point, the uterus has expanded enough to make its rhythmic surges hard to miss. Many women describe Braxton Hicks contractions as feeling like a band or a belt being tightened then released: noticeable, perhaps uncomfortable, but seldom strong enough to be painful.

Braxton Hicks was clearly fascinated by the magic of the natural world—from earthworms to women, and all species in between—and he devoted his life to the study of these marvels, not least of which was the waxing and waning of an organ that could produce human life. Science still doesn't fully understand the mechanisms that trigger full-blown labor—some potent combination, perhaps, of fetal maturity and subtle maternal signals—but the minute physiological changes that initiate true, dilatory contractions are magical enough to make even those magnificent mutton-chops stand on end. We now know that a few days prior to the onset of labor, there is an increase in the hyper-conductive gaps between the cells of the myometrium. This reconfiguration of muscle fibers allows increased electrical conduction from cell to cell until at some point—and again, we don't know exactly how or why—the uterus is afire with pulses of energy. These microscopic bolts of lightning travel across the womb in waves. We may call these waves contractions, or surges, or cramps, or simply "pains"—conflating the sensation of the thing with the thing itself—but regardless of what we call them, their purpose is universal: to bring forth new life.

"Uterine activity"—for such is the cold, clinical name given to contractions by medical textbooks—looks and feels different at various points in a woman's labor, changing according to the precise mechanism required in that moment to bring the unborn baby closer to the moment of birth. In the first stage of labor, the tightening waves of the uterus serve to nudge the fetus down into its mother's pelvis, simultaneously maneuvering the baby into an optimal position while thinning, stretching, and pulling the cervix over the baby's lowest presenting part (usually its head). As labor progresses and the cervix is open to its widest possible diameter—around 10 centimeters—contractions then become expulsive. The womb no longer simply squeezes; instead, with every surge, it forces the fetus through the pelvic outlet, and out of the vagina. This part of labor is often referred to as the "pushing" phase, and it's often depicted in film and television as a chaotic ritual in which a recumbent woman is coached through a series of bizarre countdowns and breathing sequences until gowned-and-masked medical staff finally extract a remarkably clean child from between her quivering legs. In truth, most birthing bodies push effectively and involuntarily without any coaching at all; as I often explain to the women in my care when they ask what this stage will feel like, I tell them the uterus just "throws down" in much the same way as the stomach can throw up—it is a strong, reflexive, and effortless action. The question which I have yet to answer—and the one found almost perpetually on the lips of clock-watching birth partners—is, how long does labor take? In other words, how many contractions must a person endure from that first niggly cramp to the mind-blowing moment of birth? On this subject, the uterus keeps its own counsel. Every birthing parent is unique, and so, too, is every birth, taking anywhere from a matter of minutes to seemingly interminable days. Left undisturbed, it happens when it happens, and it takes as long as it takes.

Although we may yet be unable to predict labor's onset or its duration, the human race could not have survived these past three hundred thousand years if this process had been inherently flawed. Our existence to this day has depended on a woman's ability to generate this flicker of lightning and fire it across the fibers of her womb in the right way at the right time. While many women have tragically died in childbirth—and continue to do so even in the wealthiest corners of the developed world—many more have felt the building ebb and flow of contractions in labor and have safely brought their babies earthside.

Obstetrics, a field in its infancy in the days of Braxton Hicks, has grown alongside these generations of women, becoming so much more complex and dynamic than simply "the science of standing in front." The accoucheurs of the seventeenth century would be boggled by their descendants' skill in diagnosing and managing conditions that would almost always have been fatal in times past, from preeclampsia to gestational diabetes to the most awkward and dangerous fetal presentations. But as Western medicine and its scrutiny of the uterus have advanced, so, too, has a growing sense of exasperation with the whims of the womb. We still don't know exactly why Braxton Hicks contractions occur: perhaps they are a clever self-toning mechanism that evolved to prepare the uterus for birth, but perhaps they have some other, less immediately obvious purpose.

And what of the contractions that occur too soon, some of which culminate in preterm labor, some of which twitch away for days with no obvious end result? Let's take one more moment to consider these inconveniently irritable wombs and the women who own them; let's look more closely, listen more carefully. In our rush to pathologize, we may be missing a message of life-or-death importance.

—

REBECCA FISCHBEIN SAT IN THE ER OF HER LOCAL HOSpital in Ohio and tried to make herself small. This was a tough task for someone who was nearly twenty weeks pregnant with twins: her belly was full, round, and—the reason that this was her second visit to the hospital in the space of a few weeks—it was sore. The doctor had examined her briefly, tested her urine, and decided an ultrasound wasn't necessary. The nurses bustled around the ward, just within earshot, and Rebecca knew they were talking about her.

"I was annoying," she deduced from the snippets of gossip she could hear above the beeping monitors. "I knew that I was an over-anxious mother, and I was annoying." She left the department with a lingering sense of unease, and an equally dissatisfying diagnosis: "irritable uterus." A womb that's grumpy, painful, troublesome for no reason, with a label that would almost cost her twins their lives.

Almost two hundred years before Rebecca's fruitless trip to the ER, Dr. Robert Gooch made a name for himself by inventing the diagnosis that almost killed her babies. Born at Yarmouth to a master in the British Royal Navy, Gooch progressed quickly through various apprenticeships and training in general practice before a personal crisis changed the course of his life and his career: his wife and only child died.[6] Perhaps the young medic became obsessed by the preservation of maternal life, or perhaps he just wished for a clean break from the place where tragedy had struck; for some reason—or possibly a combination of these two—Gooch moved from Croydon to central London and began to specialize as an accoucheur-physician.

The women of London provided Gooch with a rich catalog of obstetric maladies, their bodies racked with all manner of pain precipitated, in Gooch's eyes, by the mental and physical weakness particular to their sex. In 1829, the year before his death, Gooch documented his findings in what was viewed at the time as a masterful, groundbreaking text: the "Account of some of the most important Diseases peculiar to Women." This paper is praised with breathless enthusiasm on the Royal College of Physicians' website for its "pre-eminently practical character, its manly tone, devoid of trash and frippery, an ardent love of truth, a dislike of all confident assertions, and an abhorrence of all means which prostitute knowledge to notoriety or to gain."[7] What, then, could be the earth-shattering discovery described so insightfully by Gooch in this account? None other than the diagnosis given to Rebecca Fischbein and her troublesome womb two centuries later and many thousands of miles away: that of hysteralgia, or the "irritable uterus."

Gooch observed that the women of London were often debilitated by a pain originating in the uterus but often affecting, or even crippling, the entire body; unpredictable in onset, difficult to relieve, and devastating in its effect on the woman and—more important— her marriage. John G. S. Coghill, a contemporary of Gooch, summarized the condition as follows:

> *The symptoms of irritable uterus . . . may be shortly described*
> *as consisting of pain, more or less acute, in the uterine region,*
> *radiating from that centre to the lumbar and iliac regions,*
> *and extending down the thighs, often more marked on one side*
> *of the body—the left. The pain is constant, increased in the*
> *erect posture, and by any physical exertion, or even by mental*
> *emotion.*[8]

As for the womb itself, Coghill notes:

> *The uterus is found, on examination, exquisitely sensitive, and*
> *intolerant of the slightest pressure.*[9]

This statement is one of many throughout history projecting contemporary ideals of womanhood onto the uterus itself, conflating the person and her frailties with the organ and its dysfunction. Both woman and womb are seen as delicate, weak, and prone to easy upset, an attitude whose remnants are still woven through the tapestry of obstetrics to this day.

The first case study described by Gooch illustrates the peril that may befall a woman who pushes the boundaries of social conformity. Gooch's patient, a "nameless young lady" who married at the age of twenty-four and had her first child not long thereafter, committed the cardinal sin of doing too much, too soon in early motherhood. F. W. Mackenzie, another one of Gooch's contemporaries, describes this "lady" in his own writings from that time; his words paint her as a thrill-seeking vixen who springs almost instantly from childbed to party circuit:

> *After her first confinement, [she] went to a fashionable watering-place, and there passed a winter of laborious gaiety; her mornings being spent in making calls, and her evenings in standing in crowded parties. She lost her appetite, suffered much from languor, and became subject to shooting pains at the lowest part of her abdomen.*[10]

Was it the quick postpartum snapback that did her in, or the coffee mornings, or the crowded parties? We will never know precisely what caused her subsequent illness, but the message is clear—she overdid it, abandoning her maternal responsibilities for the more appealing fripperies of high society—and the treatment was as merciless as the character assassination:

> *In consequence of an accession of pain, and a sense of fulness in the womb, she applied leeches, lived low, and was confined to her sofa several weeks. At the end of this time she was supposed to be well; but subsequently she had a violent relapse, attended with great pain and tenderness across the lowest part of her abdomen. From this time she appears to have been a great sufferer, and to have undergone very heroic treatment. She was bled four times in one week . . . and mercury was given to salivation.*[11]

Nevertheless, Mackenzie writes, the patient's eventual recovery had as much to do with her modified behavior as it did with the aggressive treatment she endured, noting that:

> *All this was done with very doubtful advantage, and her*
> *eventual recovery . . . appeared to have been connected with the*
> *improvement of her general health, and a very cautious mode*
> *of living.*[12]

Other cases documented at the time tell similar tales of hysteralgic pain befalling women after various nervous upsets or bad behavior: Mrs. Woodger, whose "irritation" began after the death of her mother; Mrs. Ward, a mother of five (including one stillborn child) who had lived in "extreme anxiety" since the time of her marriage; or the singleton Miss Ross, whose "uterine disorder" was deemed a direct consequence of her work as a seamstress:

> *an occupation which is precarious and ill-requited, and which*
> *requires late hours and a sedentary mode of life.*[13]

These women come to us now as two-dimensional ghosts, with only spectral traces of the physical pain and emotional turmoil they suffered. In their rush to pathologize and classify—and in turn to be wreathed with the laurels of prestige by their peers—early obstetricians like Gooch and Mackenzie were all too keen to gather these diverse stories under the vague umbrella of the "irritable uterus." Reading between their centuries-old lines, we can speculate as to the true root cause of the women's ailments: some—having constant pain that worsened around the time of menstruation—may indeed have had what we now know as endometriosis, a debilitating and still often misdiagnosed condition. Others, like the unfortunate young socialite, may have had one of the uterine infections so common, even now, after childbirth; an illness completely unrelated to her social life or any perceived abandonment of her maternal role. Still others may have suffered simply from grief (the bereaved Mrs. Woodger) or from the cumulative stress caused by an unhappy marriage and almost relentless childbearing (Mrs. Ward), or from the exhaustion of laborious piecework (Miss Ross). In each instance, though, the maladies of the womb are blamed on the failings of the woman: an injustice that can only be corrected by a closer reading of their stories and a more nuanced scrutiny of the socio-historical context in which they unfolded.

While each case study details an individual with her own unique history and hysteralgia, together they form a mosaic of pain. One can only imagine the misery of the bereaved, the overworked, and the unhappy, and the desperation that drove them to the doors of London's hospitals. Gooch and his colleagues appear to have welcomed these women along with the opportunity to experiment with treatments that were surely as devastating as the symptoms they sought to cure. To the learned men of London's lying-in hospitals, the risk of allowing the newly minted disease of hysteralgia to progress unchecked was greater than the dangers posed by leeches and the like. Indeed, these early obstetricians believed that by treating hysteralgia, they were performing an important social function: ensuring the sexual availability and, in turn, the maternal success of the women in their care.

Robert Ferguson's preface to the 1859 edition of Gooch's seminal work explains the severe societal threat posed by a woman with an irritable uterus. The sexual dysfunction caused by hysteralgia is, in his eyes, far more dangerous than any other side effect. A painful womb may cause the adjacent vagina to become:

so exquisitely tender as to render intercourse intolerable; indeed,
in several instances, this condition has led to separation, and
the unbalanced nervous power has been the index to greater
evils, merging in insanity . . . The erotic element is in most cases
entirely extinguished. All intercourse is dreaded or loathed.[14]

Ferguson paints a startling picture of families driven to ruin by this aversion to intimacy; he writes of:

husbands estranged, children neglected, and home stripped of all
its holiest influences.[15]

In this context, the obstetrician becomes more than just a medical man; he is a potential social savior with the ability to heal vaginas, return women happily to their marital beds and thus prevent the breakdown of the sacrosanct family unit. Surely the odd side effect was tolerable in the name of pursuing this unassailably important goal?

To modern eyes, the notion of hysteralgia may be quite obviously flawed, and the associated cures—bloodletting and the like—may be relegated to the realms of quackery. It may surprise you, then, to learn that even after nearly two hundred years of medical and social progress, "irritable uterus" has been widely adopted by many practitioners of modern obstetrics as a valid diagnosis despite the absence of any concrete evidence to support its existence. Indeed, the use of the term is so prevalent that, until very recently, I myself assumed that it was a credible disorder with all of the usual regulatory guidelines for diagnosis and management. As a midwife, I've cared for countless women who have presented before pregnancy's full term with painful contractions and no associated dilatation of the cervix, and I've stood at their bedsides and watched as many doctors scratch their heads.

"You're not in labor," the woman is often told, in the absence of any clear reason for her discomfort. She is often confused—the doctor's words in direct opposition to her body's message of pain—until an alternative explanation is given with authoritative certainty: "You just have an irritable uterus." With those two words, the woman's fate is decided by those with the kind of knowledge that matters in this blue-curtained bay: there will be no drama and no baby, at least not now, not yet. Monitors are disconnected, used gloves and speculums are swept into the bin, painkillers are dispensed, and the woman is left—as common parlance would have it—to "settle."

Sure, I've often thought. Fine. I have dispensed said painkillers; I have left said woman to sit with her pain and I have moved swiftly on to the next patient, the next puzzle to be labeled, solved, and settled. But researching this book has illuminated—at least, for me—a blind spot in obstetrics' field of vision: like its illegitimate cousin, the hostile uterus, the irritable uterus barely features in today's clinical literature. The World Health Organization does mention the term briefly as a subset of "other uterine inertia" in its International Classification of Diseases, a list of billable health conditions, just below the equally nebulous and pejorative categories of "desultory labor" and "poor contractions,"[16] but a wider search reveals no definition or recognition of the term among regulatory obstetric bodies in Britain or America. Nevertheless, "irritable uterus" still hovers at the bedside—a ghostly remnant of Gooch and his peers, those early practitioners of the art of "standing in front"—its vague, shapeless form obstructing a clear view of the woman at the heart of our care.

For Rebecca Fischbein, a diagnosis of irritable uterus was almost fatally obstructive. Now an assistant professor of family and community medicine at Northeast Ohio Medical University, Rebecca knows she had all the traits that are so often protective in the implicitly biased world of healthcare: she was white, healthy, and well educated.

After absorbing the news that she was pregnant with twins, Rebecca felt ready for whatever lay ahead, confident that she would understand her options and be able to advocate for herself if need be. The first few months passed without event, but as she cruised into her second trimester, Rebecca "just didn't feel right." What began as painless, intermittent tightenings soon became regular and uncomfortable.

"I just felt like something was wrong," she recalls.

We are video-calling years after the fact, but even with the passage of time and thousands of miles between us, her frustration is fresh and clear to see. She went to the local hospital's emergency room, she tells me, and her voice is sharp with incredulity as she explains that she was given intravenous fluids and sent home. No explanation, no discussion. She felt, she tells me, like a nuisance.

As weeks passed, Rebecca's pain became more consistent and even harder to ignore, but another visit to the ER was even less satisfying than the last.

"I called again; went again," she remembers. "The same thing happened. Still no measurement, no ultrasound." At first, she felt disappointed, even dismissed, but as she sat in her bay and listened to the bustle of the ER around her, she realized that her worst suspicions were confirmed: she was actually being ridiculed. "I swear," she says, "I overheard the nurses laughing at me . . . They didn't necessarily say I was a hysterical mother, but I felt like that's what they were saying. I could hear them laughing at me like I was overreacting—freaked out for no reason." The medical staff appeared equally nonplussed by Rebecca's pain; she was sent home with a diagnosis of "irritable uterus" and advised to attend her next obstetric appointment as planned.

"I was simply a uterus that was irritable," she wrote in a later recollection of these events, "and was also irritating the healthcare professionals around me."[17]

Rebecca spent the two-week wait for that next appointment in a state of almost constant pain and worry; her "irritable uterus" was now so uncomfortable that she spent much of the time sitting or lying in bed. When the day finally came to attend her twenty-week ultrasound, the look on the sonographer's face confirmed what Rebecca's body had been telling her: something was very, very wrong with her pregnancy. One of the twins was surrounded in its sac by an excessive amount of amniotic fluid, while the other twin was small and eerily still, its own sac nearly empty of fluid and "stuck" to the wall of the uterus.

"What's interesting," Rebecca recalls, "is the ultrasound tech actually told us what she thought it was. I think they don't normally do that. We were like, what's going on? What's wrong?" As Rebecca and her husband studied the screen where their babies—girls, they now knew—floated, grainy and gray, in their mismatched sacs, the sonographer explained the urgency of the situation that had been worsening, undetected and unheeded, during Rebecca's weeks of pain.

"She was like, 'I think this is twin-to-twin transfusion. And then she went off to get the doctor." As Rebecca now knows, having spent the intervening years studying the condition and advocating for other mothers in a similar position, twin-to-twin transfusion syndrome (TTTS) occurs when a malformation exists in the vessels of a placenta shared by twins, causing one twin—the donor—to lose blood and vital nutrition, and the other—the recipient—to receive too much blood, risking cardiac overload. Both of Rebecca's girls were at imminent risk of serious illness or even death, and had been for weeks. In spite of her pain, her calls to her doctor, and her visits to the ER, no one had thought to perform an earlier scan that could have detected the condition and averted disaster; no one, it seems, had listened.

Rebecca was understandably devastated by this new diagnosis, terrified that she would lose her twins and shocked by the urgency of the technician's advice. After the ultrasound, she says, "they just put us in the hall. And I just remember bawling with all these other pregnant women walking by me. When we finally went and talked to the doctor, she basically said, 'OK, you're not going to be able to work for the rest of the pregnancy. We're going to have to refer you to the high-risk doctor [a maternal-fetal medicine specialist], and you're going to have to drive over there now, and in, like, two days you'll have to go down to Cincinnati and have surgery, and there's a good chance [the babies] are not going to survive.' I mean, it was just traumatic."

Rebecca did as she was told, reorganizing her work, attending the specialist, and undergoing emergency surgery that ultimately allowed her to carry and deliver healthy twin girls. At the time, she says, she was grateful: "I mean, I was so naive," she recalls. "I just kind of accepted it; like, OK, I guess that's what it is . . . Because without treatment, twin-to-twin is an eighty to one hundred percent fail. So yeah, it's a miracle, you know, that my girls made it." In time, though, as Rebecca reflected on the failings in her care and the near catastrophic outcome of her pregnancy, her gratitude soured. "I was so mad," she says. "I mean, the whole misdiagnosis . . . I had complained repeatedly. There were so many times they could have just done a simple ultrasound and found [the twin-to-twin transfusion] so much sooner. It took a long time to work through that anger."

Like many mothers traumatized by pregnancy and birth, Rebecca channeled her anger into activism, devoting years of research to understanding women's experiences of TTTS and helping others navigate the system that let her down. In one major study of 367 other women with TTTS, Rebecca and her coauthors found that just over half of mothers who had symptoms and shared them with their providers felt that their complaints had been dismissed.

"The point of my research," she says, "is that the patient's voice is key. We have to advocate for ourselves. Women know their bodies best. If a woman feels something's wrong, something's probably wrong. There's such a power difference between the patient and the provider, and we tend to just accept what they say, like, 'OK, they're the expert, they know.' But you just have to keep fighting."

Admittedly, not all women given the misdiagnosis of "irritable uterus" will have life-threatening illnesses. Some will have nothing more serious than a gastric upset or urinary tract infection, both known to cause cramping in pregnancy; others will, indeed, "settle," with no cause found and no consequence to suffer. However, Rebecca's story illustrates the dangers of absorbing terms like "irritable uterus" into common obstetric parlance. Invented by men keen to make their own stamp on an emerging profession and originally used as a catchall term for a catalog of psychosomatic symptoms, the term and its application have evolved beyond anything Gooch and his contemporaries could have imagined. With no clear definition, and a course of management often guided by guesswork, "irritable uterus" is a dangerous misnomer. As such, the term distracts from the root causes of a birthing person's pain and diminishes their embodied knowledge. Women like Rebecca—and like Mrs. Ward, and Mrs. Woodger, and the unnamed socialite whose "winter of laborious gaiety" was her downfall—have long been dismissed. We may think we've traveled light-years from the nineteenth-century lying-in hospitals and the antiquated attitudes of those who worked there, but by continuing to conflate a woman and her uterus into one troublesome package—both irritable and irritating—we have advanced little further than our forebears. We want the womb to behave in a specific, prescribed, industrially convenient way; we write many-paged, multi-bullet-pointed guidelines to that effect, and we work within their narrow goalposts. We "stand in front," as the original meaning of obstetrics would have it, but as Rebecca Fischbein's story illustrates, we sometimes stop seeing and hearing what is most important.

Labor

OXYTOCIN AND THE GOLDILOCKS CONTRACTION

The hurrier I go, the behinder I get.

—AMISH SAYING, SOURCE UNKNOWN

The year is 2011, and I am a student midwife in the labor ward. The woman on the bed in front of me is having a heated argument with her partner about his inability to pack the correct bra in her hospital bag, but their squabbles fade to white noise. The only voice I hear—the only one that matters to me in this moment—is that of my mentor, Betty,[1] a senior midwife with a no-nonsense attitude and a diminishing tolerance for my ineptitude. Tonight, after weeks of assessing my skills and my mettle, Betty is showing me how to perform one of modern obstetrics' most important tasks. Not catching a baby—that came within minutes of my first-ever shift the previous year, my hands scrabbling to grasp the slippery bullet of limbs and slime that shot toward me with alarming speed. Not suturing a tear—that will come later, my mind boggling at the shock and thrill of pulling a hooked needle so easily through the tenderest flesh of another human being. No, tonight—at 4 a.m., that witching hour when fatigue makes every movement seem both vivid and unreal—I am learning how to perform a special, secret kind of sorcery: induction of labor.

The woman on the bed is more exhausted than I am; her baby is nearly two weeks overdue, and she's been pacing the floor of the antenatal ward for days, cramps rolling through her belly while she waits for the magical alignment of stars, staffing, and bed space that will herald her admission to the labor ward. Finally, only moments after falling into a fitful doze, she felt a hand shaking her shoulder and heard a voice telling her, "They can take you now." Now, still rubbing the sleep from her eyes, and wearing a thin cotton gown that says "HOSPITAL PROPERTY" in a million tiny letters across her body, she is here. She has "made it," as Betty has announced with dry, faux pomp. And I am tasked with making "it" happen.

Like chefs preparing for an elaborate meal, my mentor and I gather our ingredients from cupboards, drawers, and fridges around the ward before setting them out in careful piles on the metal cart in our room. There is the half-liter bag of electrolyte solution, the cannula that will be slipped into the woman's vein, and the coiled plastic tubing that will connect one with the other. There are needles and syringes, brightly hued and plastic-wrapped like candies; we will use them to draw up and infuse the drug that sits at the edge of the cart in a tiny glass ampoule no bigger than the tip of my thumb.

"Jungle juice," Betty says. She raises the ampoule up to catch the cold light of the examination lamp extending its long metal arm from the ceiling. There is only one milliliter of fluid inside the little bottle; my mentor tilts it to and fro, swirling the liquid against the glass like a proud sommelier. "The good stuff," she announces, showing me the label: "Syntocinon, 10 international units in 1 milliliter." In the next moment, I sense a shift of intention—a set to her jaw, a barely perceptible draft in the room. She flicks her nails once, twice, three times against a tiny white dot on the neck of the bottle, grasps the tip of the ampoule between her thumb and forefinger, and snaps its top clean off. Betty allows a second of stillness as the snap echoes around the room, and then her movements are quick again; practiced and purposeful.

I am supposed to be paying attention, watching carefully as she draws the Syntocinon into a syringe before injecting it into the port of the bigger bag of fluids. She is saying something about changing needles, about holding the bag flat and piercing it at a certain angle so we don't stab ourselves. This is important, I know—keeping our own bodies intact while controlling the body of another—but all I can think about is the dusting of fine glass powder in the creases of her hand. In this liminal hour between night and day, gestation and birth, Betty is a dark queen, glistening with fairy dust, iridescent as she goes about her alchemy. I am wide-eyed and cotton-mouthed at her side. We are about to make something happen, and in the years to come, I will make that same thing happen over and over again for hundreds and hundreds of women. I will coax sluggish wombs until they contract strong, fast, and hard. I will make them labor. I will still wonder, in the years to come, if this is magic or madness. The question echoes with each snap of glass in my hands.

–

ROBERT GOOCH, JOHN BRAXTON HICKS, AND REBECCA Fischbein have shown us the perils, peculiarities—and sometimes perfectly normal qualities—of wombs that tighten and surge before pregnancy's full term has been reached. What, though, of the contractions that are stubbornly slow in coming, even when a pregnancy has continued well past term, frustrating the would-be mother and turning her caregivers into clock-watchers? What, even, of the contractions that slow or stop in the middle of a seemingly straightforward labor? Medicine has long responded to these stuttering stops and false starts with a kind of obstetric two-step: frustration, followed swiftly by intervention.

From the "lying-in hospitals" of Victorian London to today's modern labor wards, doctors and midwives have used all methods mechanical and pharmacological in their hunt for the holy grail: the Goldilocks contraction, neither too slow nor too quick, not too early nor too late. Like that final bowl of the bears' porridge in the fairy tale, the Goldilocks contraction is "just right," its presence securing the safety of mother and child while also suiting the industrialized time scale of the modern maternity unit. The Goldilocks contraction fits both the mother and the machine in which she labors.

As a midwife in a busy urban hospital, I, too, have been taught to chase this goal. Women with high-risk pregnancies and complicated histories come to us for care, but as so often happens, those fit, healthy people who fall into obstetrics' increasingly slim definition of "normal" soon become swept up in the tsunami of intervention once they've crossed our linoleum-tiled threshold. Although I began my training as a devout believer in the power and wisdom of birthing bodies— and I like to think I still hold those values fervently in my heart—the pressures of environment and culture have made me complicit in this often futile search for the "perfect" contraction. Under the guidance of Betty and countless other mentors and colleagues—most of whom mean well, but all of whom must work within a regimented, risk-averse, large-scale system—I've scrutinized, monitored, welcomed, and cursed capricious wombs in a million different ways. I've palpated "uterine activity," as we so coldly call it, by resting my hand on the gathering tension of an abdomen made numb by epidural anesthesia. I've watched ghostly green numbers rise and fall on the screen of a bedside monitor. I've seen women's faces twist from smile to grimace as their bodies are gripped in the clutch of a million fibers tensing as one. I've heard laborers hoo and haa and praise God and curse their bodies and their husbands, and I've listened for the telltale grunting sigh that signals the beginnings of a push. I've marveled at the beauty in these sights and sounds, but I'll admit that I've all too often lost faith in the womb in front of me. While my intentions have always been good and my prayer always for a swift, safe, and joyful birth, I'll admit to being swayed over time by those around me, to saying, sotto voce, "She's not up to much," or "She's making heavy weather." It is a short but terrifying leap, I know, from these thoughts to the one voiced by an obstetrician to a midwife friend of mine. "Some women's wombs are just crap," he said, a remark meant to console during a labor that refused to progress in spite of everyone's best efforts, but whose sting is inescapable and, some might say, inexcusable.

AND SO, TO CAJOLE THESE WOMBS INTO ACTION— whether to optimize outcomes for mother and child, or to fit the system's schedule, or some muddy confluence of both—we snap that magic glass and begin the process of induction of labor. Every day, in hospitals around the world, thousands of midwives like Betty and me make their preparations: drawing up a syringe of synthetic oxytocin, infusing it in a bag of fluid, and running it through a pump that sends powerful pulses of hormone—first just a few drops an hour, then more, then more—into a birthing body's bloodstream. Often called the "hormone of love," oxytocin is indeed released during moments of intimacy, including infatuation and orgasm. Perhaps most important, though, oxytocin is responsible for initiating and sustaining contractions of the uterus—expelling both the fetus and its placenta—and is released in great quantities during the first moments after birth, encouraging a period of bonding so important and intense that it is now often referred to by midwives as "the golden hour."

Reasons for induction of labor are almost as numerous as the bodies themselves; in many cases, the procedure is offered because the baby is overdue or "post-dates," although due dates themselves are not an exact science and the tipping point of fetal maturity is subjective. Some research demonstrates an increasing risk of placental insufficiency—a dangerously diminishing supply of blood, oxygen, and nutrients from mother to fetus—as the pregnancy approaches or passes two weeks beyond the estimated due date. These data have led to widely varying induction protocols: some countries have adopted routine induction of labor at thirty-nine weeks (one week before the due date), ostensibly to avoid any risk whatsoever of postmaturity. Many practitioners wait until forty, forty-one, forty-two weeks or even beyond, whether to give the best chance of spontaneous labor, or to spare a hospital's bed state, or both. Induction may also be suggested in any other situation in which the fetus would appear to be better off out than in; for example, if there is evidence that a baby's growth has begun to slow or "tail off" or, conversely, if the baby seems to have grown so large that waiting any longer for spontaneous labor to begin might risk injury to mother and child. Reduced fetal movements are another common reason for induction, as fewer kicks and rolls sometimes indicate fetal compromise; reduced amniotic fluid is another.

In recent years, reasons for induction appear to have multiplied even more: in addition to the common but opposite reasons of "small baby" or "big baby," induction may be routinely offered if the pregnancy is a product of assisted conception, if the mother is over a certain (somewhat arbitrary and varying) age, if she is diabetic, or if a condition such as severe pelvic pain has made continuation of the pregnancy past full term unbearably uncomfortable. Increasingly, "social" or elective induction is also offered, with parental preference (and provider availability) dictating the date when labor will begin.

As the list of widely accepted reasons for induction has become longer and more creative, the number of birthing people undergoing this process has grown exponentially. From 2020 to 2021, roughly one third of registered births in America and England were induced.[2,3] It's important to note that these figures do not include the number of labors that began spontaneously but were then "augmented," or accelerated, with synthetic oxytocin—a process that often happens if contractions appear to have slowed or stalled, especially in the latter stages of labor. Statistics for this kind of intervention appear patchy across the globe, perhaps reflecting the fact that "helping" labor in this way is so commonplace as to be unremarkable and therefore unworthy of documentation. Regardless of the reason for induction or augmentation, the mechanism remains substantially the same: synthetic oxytocin—the hormone that initiates contractions—is administered intravenously in ever-increasing doses. Ready or not, the dormant or sluggish uterus is roused to life: first niggling, aching, and cramping, then surging and heaving, and finally pushing as billions of muscle fibers fire in unison toward the explosive finale of birth.

Even after pharmacologically initiated or assisted births have been considered, the number of women receiving synthetic oxytocin is undoubtedly much higher. Not only is the drug widely used in labor, it is almost universally offered throughout the developed and developing world to expedite delivery of the placenta and minimize postnatal bleeding. In an unmedicated third and final stage of childbirth, the uterus initiates a complex but effective chain of events after the expulsion of the fetus: the muscle continues to contract, but in a way which prompts the vessels in the womb's lining to effectively tie themselves off. The action of these "living ligatures," as they are often called, encourages the placenta to shear away from its site with a physiologically tolerable amount of blood loss (usually around 500 milliliters or less). When successful, this process generally unfolds in less than an hour, and is aided only by "maternal effort": a spontaneous sensation of pressure that encourages the woman to bear down and expel these last remnants of pregnancy.

With the advent of synthetic oxytocin, though, a new protocol has emerged which is now dominant across much of the birthing world: at or around the moment the baby is born, synthetic oxytocin is given to the mother as an intramuscular injection, the umbilical cord is clamped, and the placenta and membranes are delivered, often in five to ten minutes, by steady, firm traction on the cord.[4] This regime, known as "active management" of the third and final stage of childbirth, appears to have been so widely adopted as a routine intervention that it, too, is absent from systematic documentation in official state registers. However, in one UK survey of over four thousand birth professionals, 93 percent of obstetricians and 73 percent of midwives reported "always or usually" practicing active management.[5]

Stultifyingly dull as it may be to trawl through birth census figures, I include these data here to demonstrate just how prevalent the use of synthetic hormones in labor has become. Let's reframe this simply in terms of the uterus: around 30 percent of wombs across the Western world are nudged into labor by synthetic oxytocin; a further, unknown number are given the hormone to encourage longer, stronger, more regular contractions once labor has begun; and an undetermined but almost certainly much larger number are given oxytocin to ensure a quicker, cleaner, and (in some, but not all, circumstances) safer delivery of the placenta and its membranes. In short, if you are a birthing person in the twenty-first century, there is a strong chance that at some point in the beginning, middle, or end of your labor, you will be told that your uterus is not enough—that its efforts are too feeble, too irregular, or too dangerous. Too little, as the saying goes, too late.

-

HOW DID WE GET HERE? THE JOURNEY BEGAN MANY years ago and has been a relay of countless sprinters and plodders, each one passing the baton of obstetric progress to the next—some more skillfully than the others. For millennia, midwives and medics have been nudging wombs toward the finish line, whether to expedite labor or to induce abortion. The latter procedure has been an integral element of reproductive healthcare and self-care since the earliest days of recorded civilization; the basic human need to manage one's fertility and discontinue unwanted or life-threatening pregnancies is evidenced by descriptions of abortifacients in ancient Egyptian, Chinese, and Roman texts.[6] Dioscorides, a physician and surgeon in first century AD Greece, praised the contractile effect of the cyclamen plant; according to a later translation of his work, "They say that if a woman great with child doe goe over ye roote, that shee doth make abortion, and being tyed about her it doth hasten the birthe."[7] Over the following millennia, birth and abortion were "hastened" using whatever natural materials were most readily available and effective. The European pharmacopeia included pennyroyal, rue, wormwood, and sage, while enslaved African and Native American women used a number of remedies, including those derived from cotton root[8] and the seeds of the peacock flower, *Flos pavonis*,[9] to coax and control their wombs in circumstances which were otherwise uncontrollable. Emily West, professor of history at the University of Reading, frames the use of abortifacient plants as a means of physical, moral, and commercial resistance:

*Some women resisted slavery by working to control their
participation in producing more enslaved children. New mothers
risked eventual separation from their children, and if they had
daughters, they knew any sexual violence they had experienced
could become their daughter's experience as well. Women used this
kind of resistance, therefore, to combat slaveholders' control over
their own bodies and protect their potential children from the
horrors of bondage.*[10]

The depth and breadth of accumulated expertise in this subject
cannot be underestimated: in every continent, in every century—and
often in the face of unthinkable oppression—pregnant people and their
caregivers have devised and shared recipes to initiate and sustain the
womb's contractions. In many instances, the desired action of these
tinctures and potions also came with unpleasant or downright dangerous side effects: never was this truer than in the case of ergot, a substance whose properties were discovered by unhappy accident in the
Middle Ages, and later refined to become one of the most widely used
uterotonics of modern times.

A fungal growth found on sheaves of rye, ergot was a substance of
plentiful availability and potentially devastating toxicity. The physical effects of ergot consumption were initially observed in those who
had eaten bread made with contaminated rye flour: venous contractions soon caused twitches, spasms, and ferocious burning in the
hands and feet, giving ergotism its commonly known moniker of
"St Anthony's Fire."[11] Manuscripts written by male physicians in the
fifteenth and sixteenth centuries suggest that midwives, having observed the strong contractile properties of ergot, soon began to use it
to initiate and enhance labor for both live and abortive births. Three
different German scholars of the time described prevalent use of ergot for obstetric problems; tinctures and powders were prepared for

permutter and *heffmutter* (uterine pains), and for treatment of postpartum hemorrhage.[12,13,14] Careful preparation of ergot appears to have been widespread in France, too, by the eighteenth century; in 1774, a letter from French midwife Madame Dupille describes the administration of thimblefuls of diluted ergot for augmentation of labor[15] and physician Jean-Baptiste Desgranges observed midwives in Lyon using powdered ergot for the same purpose.[16]

The potent fungus that caused St Anthony's Fire soon set the world of American medicine alight, too. In 1807, New York physician John Stearns wrote a letter to a colleague in which he described the powerful effects of ergot with the kind of breathless enthusiasm you might expect of any modern-day pharmaceutical rep. Stearns, a Yale-educated doctor whose dedication to clear-eyed, evidence-based medicine prompted him to cofound an anti-quackery society, admitted that he learned about ergot from "an ignorant Scottish midwife." Professional epithets aside, his admiration for the substance is clear:

> *It expedites lingering parturition, and saves to the accoucheur a considerable portion of time, without producing any bad effects on the patient. The cases in which I have generally found this powder to be useful, are when the pains are lingering, have wholly subsided, or are in any way incompetent to exclude the foetus . . . The pains induced by it are peculiarly forcing . . . In most cases you will be surprized with the suddenness of its operation; it is, therefore, necessary to be completely ready before you give the medicine, as the urgency of the pains will allow you but a short time afterwards. Since I have adopted the use of this powder I have seldom found a case that detained me more than three hours.*[17]

It is interesting that Stearns, who writes about a midwife in such derogatory terms, was so quick to adopt one of her remedies into his own obstetric arsenal. In fact, evidence from excavations of Soutra Aisle, the site of an ancient monastic hospital in Scotland, suggests that the use of ergot in the country may date back as far as the twelfth century.[18] According to Dr. Rupa Marya and Raj Patel, authors of *Inflamed: Deep Medicine and the Anatomy of Injustice*, this is a recurring trope in the history of Western medicine: women's wisdom is simultaneously derided and co-opted by male medics seeking to promote their own careers. They write, "Women's medical knowledge has been stolen, and women have been used as laboratories for domination."[19] Nowhere can this be observed more clearly than in the quest for control of the uterus and its behavior in childbirth.

To Stearns, the womb was there to be dominated, and the quicker, the better. He appears to have been captivated by ergot's potential to make births faster and more efficient than ever before, with the three-hour time limit clearly extolled as a unique selling point. One can only wonder how women felt about this new kind of "forcing," "sudden," and "urgent" labor; their opinions (and any finer points of ergot use that may have been mentioned by the "ignorant" Scottish midwife) do not merit inclusion in the doctor's letter. As for Stearns's preference for ergot over the other uterotonic and abortifacient remedies already so prevalent in the United States—for example, the cotton root used by enslaved Black women of the American South—this, too, remains a mystery. Perhaps the dominance of ergot is emblematic of the culture in which it emerged: a white patriarchy with little respect for the embodied wisdom of women, especially women of color. This, then, was the historical backdrop against which modern induction of labor was born. Stearns's notes on what he called *pulvis parturiens*—"Birth powder"—appeared in the *Medical Repository* in 1808,[20] and with their publication, childbirth in America—and, ultimately, the rest of the world—was changed forever. Doctors were no longer at the mercy of the uterus, that most erratic and unpredictable of organs. Stearns and his colleagues welcomed the dawn of a new age of obstetrics—one that persists to this day—in which speed and efficiency are prized above patience, comfort, and respect for the physiological process of labor. It seemed as if "lingering parturition" was a problem for which there was, at last, a solution. The Goldilocks contraction—neither too early nor too late; not too fast nor too slow—was within tantalizing reach.

Unfortunately, women on the receiving end of this new kind of obstetric "care" soon began to suffer the ill effects of often indiscriminate and excessive use of ergot. In their haste to administer the drug, some physicians abandoned the cautious and undoubtedly safer "thimbleful" approach that appears to have been practiced and refined by European midwives in the preceding centuries. Doctors began to observe ergot's less desirable effects, from vomiting and high blood pressure to hypertonic contractions—powerful spasms of the womb that can cause injury or death to both mother and child. David Hossack, a contemporary of Stearns, and founder of New York's first Lying-In Hospital, was at the forefront of this backlash. Observing ergot's contribution to an alarming increase in stillbirths, Hossack quipped that the drug should be more aptly renamed *pulvis ad mortem*—powder of death.[21]

Ergot's increasingly negative reputation as a blunt and sometimes harmful instrument persisted, and clinicians increasingly—although not always safely or successfully—turned to alternative methods of induction. Quinine, castor oil, douches, and enemas gained favor in various circles, as did mechanical methods of dilating the neck of the womb. Some medics favored inflatable bags or catheters that could be passed into the uterus; others preferred expandable instruments like Dr. James Simpson's "Sea-Tangle Tent"—a stalk of dried seaweed which, once inserted, opened the cervix gradually as it absorbed moisture from the surrounding tissues.[22]

Such makeshift and often risky techniques persisted until a

remarkable discovery at London's University College Hospital in 1935. John Chassar Moir, an obstetrician, and Harold Ward Dudley, a biochemist, isolated the active ingredient in ergot and produced it in a form that could be administered relatively safely as an intravenous or intramuscular injection. This new drug, which the pair dubbed "ergometrine," was given to women in their first week after birth in the hope of minimizing or preventing postpartum hemorrhage, a common but potentially fatal event in which insufficient contractions of the uterus allow the womb's highly vascular lining to continue to bleed catastrophically in the hours, days, or even weeks after birth.

Moir and Dudley noted that careful administration of ergometrine appeared to produce strong, regular uterine contractions without any of the unpleasant effects of other, less refined forms of ergot:

> *Whereas [other preparations], when given in a dosage large enough to produce a definite action on the uterus . . . produce in the patient a feeling of depression, headache, nausea and even vomiting, the new substance in useful clinical dosage is remarkably free from such side effects. This has often been proved by seeing the patient eat a large lunch or fall asleep after the administration of the test dose.*[23]

One can imagine Moir and Dudley standing triumphantly at the head of a ward full of dozing women, starched sheets stretched across bellies full of steamed puddings and ergometrine. This vision of passive, placid bodies seems to fill the authors with paternalistic pride; the ward of napping women is presented as almost as much of an accomplishment as the prevention of hemorrhage itself. It is tempting to tut and sigh at yet another patriarchal conquest of the birthing landscape; however, the advent of ergometrine has a bittersweet twist in its tale. Moir altruistically sought no patent or profit from his discovery, insisting instead that ergometrine's formula should be freely and publicly available for the benefit of women everywhere, and Dudley died on the day of the study's publication.

While Moir and Dudley were busy taming the wombs of London's women, a young scientist called Vincent du Vigneaud was just beginning to make some even bigger waves across the Atlantic in America. Born in Chicago, Vincent du Vigneaud—or "VdV," as his colleagues later named him—had an early aptitude for invention and efficiency; in high school, he was a prolific maker of homemade explosives using materials bought from the local pharmacist, and in the summer after his senior year, young VdV joined a wartime scheme to work on one of the many farms outside the city, where he discovered a natural ability to milk twenty cows by hand in a single session. Subsequent jobs as a soda jerk and apple picker helped finance a chemistry degree, and so VdV began a steady rise through the field, first achieving his PhD, later becoming a departmental chair at Cornell University, and finally achieving worldwide adulation for his work on sulfurous compounds including insulin and oxytocin, the hormone of love and labor.[24]

The existence of this special hormone had been discovered several years before du Vigneaud was born. In 1909, Sir Henry Dale, an English physiologist and pharmacologist, found that an extract from the posterior pituitary gland could initiate contractions in the uteruses of pregnant cats.[25] Dale named the substance oxytocin—meaning "quick birth" in Greek—and subsequent experiments on both sides of the Atlantic confirmed the hormone's contractile effects. Guinea pigs, cats, rabbits, and dogs were among the first creatures to be induced in this way, and some researchers began tentatively to trial these new methods on women, too. In his 1942 doctoral thesis for the University of Glasgow, medical student George Howard Bell bemoaned the difficulties of translating animal studies of oxytocin into human success:

It may perhaps be called an academic piece of investigation, but in the doing of it I have at times fought with recalcitrant cows brought into the byre in the middle of the summer rains and at other times I have auscultated fetal hearts in the midst of the startling whiteness of a modern maternity hospital.[26]

The latter setting and the patients within it provided their own unique challenges; while "recalcitrant cows" might eventually be wrangled into submission, human females—often subjected to hormonal experiments without analgesia—experienced side effects such as faintness, nausea, pain, and even, as in one trial from 1940, "a feeling of suffocation."[27] Oxytocin could clearly do wondrous things, but isolating it was difficult, refining its use across species was challenging, and synthesizing it in a mass-producible form remained a pipe dream.

Enter bovine enthusiast and keen chemist Vincent du Vigneaud. VdV may not have been particularly interested in the intricacies of childbirth or in the touchy-feely moments of the postnatal golden hour, but he was fascinated by the hormone behind these phenomena. In 1955, he won the Nobel Prize in Chemistry for isolating oxytocin, determining its chemical composition, and—in a major milestone for modern obstetrics—synthesizing the hormone for the first time, thus enabling its mass production and widespread pharmacological use. Commercial patents soon followed.

Somewhat fortuitously for du Vigneaud, his discovery came at a time when an optimistic postwar world was keen to embrace any new technology or invention that carried even the slightest whiff of futuristic promise. After years of deprivation and struggle, the 1950s ushered in an age of joy and exploration; families flocked around the first color televisions while bobby-socked teens jived to long-playing records and the Soviets launched their satellite, Sputnik, into orbit. Dreams became reality; the impossible became possible. Medicine was hardly immune to the allure of space-age progress. If Russians could send a rocket around the Earth, what similar wonders could be achieved on the frontiers of "inner space" within the human body?

In a surge of warp-speed progress that would have boggled young VdV's mind back in his soda-jerking days, obstetricians around the world seized on synthetic oxytocin as the drug that could launch obstetrics into a new, higher, faster orbit. Within months of publication of his initial findings in 1953, and well before the drug was licensed in America or abroad, medics had already begun to test the effects of synthetic oxytocin on the uterus. At the University of Pennsylvania Hospital, obstetrician Edward Bishop was busy trialing an exciting new procedure called "elective induction of labor," in which he used varying sequences of oxytocin administration and artificial rupture of membranes (breaking women's waters, in common parlance) to initiate and accelerate labor. After performing one thousand such inductions, Bishop decided that four hours was the "optimal" duration of labor[28]—an echo of John Stearns's three-hour bedside visits, and a surprise, perhaps, to anyone who has observed the somewhat longer average duration of a spontaneous, unmedicated birth. As in Stearns's time, labor was reframed as a problem to be solved, only this time, the space-age aspirations of the 1950s reimagined the uterus as a machine that could be manipulated to perform within strict parameters of efficiency and expectation.

To reinforce this model, Bishop even created a scoring system that allocated "points" to a woman's body according to, among other things, the ripeness of her cervix; even today, this system—Bishop's Score—is widely used to determine how "favorable" a uterus may be for the onset of labor. In the same year, Emanuel Friedman, an obstetrician from New York, used a study of five hundred women—some of whom were given synthetic oxytocin—to determine the average rate of cervical dilatation.[29] The resultant graph—Friedman's Curve—is, like, Bishop's Score, still used to guide management protocols and guidelines in modern birth settings around the world. These guidelines remain widely regarded as obstetric gospel despite the original studies' relatively small sample sizes, and the proliferation of subsequent research indicating (as mothers and midwives have long known) that the rate of dilatation in labor can vary widely and still result in safe, successful birth.

The clear, constant seam running through this new mechanistic model was VdV's wonder drug—synthetic oxytocin—and the rush to license the substance was almost as hasty as the rush to expedite labor itself. From 1955 to 1956, this man-made hormone was licensed first in the USA by Parke-Davis as Pitocin, then in Europe by Sandoz as Syntocinon—both brand names by which the drug is still known today. Thousands of little glass ampoules containing du Vigneaud's precious liquid soon rolled off the assembly line, into the hands of obstetricians and onward, through the veins and the wombs of laboring women around the world. The drug was enthusiastically promoted by manufacturers and warmly welcomed by medics as a cutting-edge solution to the age-old "problem" of the birthing body, with its unpredictable, sometimes slow, progress, and its need for constant care and attention.

An educational film written by and featuring British obstetrician D. J. Macrae in 1949 presents chemically induced labor as a clean, clinical, and virtually contact-free procedure. Macrae narrates as "Sister"—a pretty nurse in starched cap and apron—attends to her "patient": a woman tucked so tightly into her bedsheets that only her perfectly coiffed head is visible, disembodied and stark, against the crisp white field of her pillow. Sister adjusts the hormone drip flowing from a glass bottle at the woman's side before drawing back the bedsheets to reveal another recent invention: a "sonoscope" or microphone strapped to the woman's abdomen and connected to an amplifier in the next room. "Sister makes the patient comfortable," Macrae drones in the clipped, flat tones of received pronunciation. "She returns to her office to continue with her charts and, switching on the cardiophone [the amplifier], keeps continuous and reassuring watch on the fetal heart, the fetal movements, and the uterine contractions."[30]

This technologized, depersonalized model of care, in which the metrics of labor are monitored remotely and the birthing attendant is expected to focus more on her administrative tasks than on the mess, heat, and blood of the birthing body itself, is one which many modern midwives—myself included—will recognize. The scene in which "Sister" supervises labor from afar provides the blueprint for today's industrialized maternity units, in which staff monitor digitized readouts of the fetal heart and uterine activity from a central bank of screens. In these settings, the womb itself is a distant, unseen thing—controllable, quantifiable, and clean.

Vincent du Vigneaud could not have anticipated the effect his invention would have on the birthing world. As induction of labor evolved from lab-based theory to accessible, affordable practice, the gospel of Syntocinon spread rapidly around the world. One 1959 advertisement from the pharmaceutical company Sandoz announced "the first industrial realization of synthesized oxytocin" next to an image of a glass ampoule pointing toward a disembodied uterus.[31] The message could not have been clearer: chemicals can do to the womb what nature cannot muster.

This approach appears to have been successful in capturing the collective obstetric imagination: within a few years, synthetic oxytocin was so widely used that some medics began to refer to the drug simply as "normal saline" (which is, in fact, a completely different and essentially benign solution of mildly salty water used for rehydration). By the 1970s, one paper from Brazil—where induction of labor had been given the somewhat terrifying but accurate name of "narco-acceleration"—expressed the commonly held view that "the obstetrician's interference in labor today is an obligation, and even more, a duty. Childbirth cannot and should not progress without the guidance of an obstetrician attempting to lessen pain, shorten labor, correct abnormalities, and provide support and psychological assistance to laboring mothers."[32]

Decades later, this approach still prevails, positing obstetricians as heroic saviors, riding to the rescue of women who otherwise would have been held hostage by the whimsies of their wayward wombs. Australian midwife Rachel Reed argues that this narco-accelerative model is inaccurate at best, and dangerous at worst. "The only accurate indication of an effective contraction pattern is the birth of a baby," she writes in her manifesto, *Reclaiming Childbirth as a Rite of Passage*. "The idea that contractions must fit particular criteria to be effective conflicts with the reality of women's unique contraction patterns. I have witnessed many women birth babies perfectly well with very irregular and spaced-out contraction patterns. According to the prescribed criteria, these women were never in 'established labor.'"[33] Elizabeth Newnham, an Irish midwife, contrasts these "irregular" laboring behaviors with an obstetrically led system in which "institutional momentum" is a key driver of policies and protocols. In an ethnographic study from 2017, Newnham and her coauthors describe labor care characterized by anxiety and clock-watching; institutional momentum drives the study's interviewees—midwives and doctors—to question physiological variations and initiate frequent interventions. The authors argue that "the institution imposes an externalized and artificial timeline on a process that is individually and uniquely experienced," and "Synt"—Syntocinon—is frequently cited by midwives in the study as a panacea for labors that diverge from this timeline.[34] Induction and augmentation keep the momentum going, for better or for worse.

It would be churlish, though, to argue that institutional momentum is a product of institutional malice. Newnham and her coauthors accept that this impetus to accelerate and intervene is driven by a desire to minimize or mitigate risk, and as a midwife working in exactly this kind of risk-averse institution, I can confirm that anxiety—however justified or misguided—around the safe, timely birth of a healthy baby is, by far, the primary reason for using synthetic oxytocin. Many women may rightly be grateful for a drug that enables delivery of babies who otherwise would have been compromised by remaining *in utero*. Induction of labor may unequivocally be the safest strategy for many pregnancies; however, for many others, the exact balance of risks versus benefits remains unclear. A 2020 review of thirty-four randomized controlled trials including over twenty-one thousand women "at low risk for complications" and their infants found that, compared with "expectant management" (i.e., waiting for labor to begin spontaneously or waiting for more time to begin induction), induction of labor at or beyond thirty-seven weeks of pregnancy appeared to be associated with fewer deaths of babies around the time of birth. It should be noted that the overall number of deaths in each group remained small, relative to the large sample size (four perinatal deaths in the induced group versus twenty-five such deaths in those who were managed expectantly).[35] More recently, however, a 2021 study of 474,652 births found that babies born following induction of labor experienced more birth trauma, were more likely to need neonatal resuscitation, and to require hospital admission for respiratory issues even up to the age of sixteen. The incidence of adverse maternal and neonatal outcomes was even higher in those inductions carried out for nonmedical reasons: a not-insignificant 69,397 births, or roughly 15 percent of the whole.[36] The jury, then, may yet be out: while induction of labor may save certain infants from imminent danger or death, it is sometimes unclear at the outset who those infants might be and how best to help them. Further research and more nuanced analysis of conflicting evidence are needed to determine why, how, and when to intervene.

Some critics of the widespread—and arguably disproportionate—use of induction of labor make the point that neonatal outcomes should not be the definitive measure of the procedure's merits. While the safe delivery of a live child is obviously of paramount importance, many birthing people and their advocates suggest that this is not the only indicator of a "good" birth. Labor and birth represent a truly formative moment in many women's lives, and as such, they have long-lasting impacts on physical and mental health that should be carefully weighed alongside the well-being of the child.

Some healthcare providers may suggest that a healthy baby is an unassailable reason for induction of labor. That may be true in many cases; for example, journalist Jennie Agg describes her openness to the prospect of induction after suffering four miscarriages in the preceding years: "I was afraid of my baby dying. More than anything. So what then? What counts as a positive birth under those terms?"[37] For her, induction—a concept she'd previously seen as "the thing you're not supposed to want"—offered the prospect of a live baby at the end of an emotionally exhausting pregnancy. Other women, though, tell a story of consenting unquestioningly to induction, only to find that the experience leaves them with an enduring feeling of disconnection or even trauma.

Having given birth to two healthy children—one induced labor, and one spontaneous—journalist Alex Beard knows that her connection with her body and her faith in its ability matter, too. Speaking to me via video link from her makeshift recording studio—a cupboard piled high with her family's coats and sweaters—Alex recalls the emotional fallout from the birth of her first child. Faced with the possibility of induction after her waters broke without any accompanying contractions, she describes her increasing anxiety as time passed without any obvious signs of labor. "I had a feeling that hopefully [my contractions] will kick in," she tells me. "They *should* kick in. And every day I woke up and nothing was happening, I felt like a bit of a failure—just forty-eight hours of feeling like a failure with this induction impending." Finally, having been warned of the small but increasing risk of infection with every passing day, Alex went to the hospital to be induced, but even as the midwives hung "bag after bag" of Syntocinon infusions, she still waited anxiously for her body to perform as expected.

"It was a full day," she recalls. "I knew I had the Big Boy Drugs in me," she says, "so I felt like I was kind of hurtling toward this edge of a cliff at which point I would lose all control. That's kind of where I found myself."

Over many following hours, Alex waited for the hormone to encourage her body's own natural contractions, as she had been told it would, but reality was far less straightforward.

"Any time the drip finished, the contractions stopped. At no point during that labor did my body kick into natural labor. At no point did those drugs kick-start the mechanisms that it should have. It was just bag after bag after bag at that point," she recalls, until suddenly, as so many women in my own care have remarked, contractions began with a ferocity and frequency for which Alex was hardly prepared.

"It was, 'Boom, this is happening,'" she remembers. "I thought, well, the contractions are coming this thick and this fast, so my cervix is going to be opening. And I will never forget the woman who came and did an internal examination and said, 'You're two centimeters [dilated],' and my heart just absolutely sank because I was like, we're hours away from anything."

"I was exhausted," Alex tells me. It is clearly difficult for her to recall this day of seemingly endless pain and frustration; it is diffi-cult, too, to listen. "The contractions kept coming thick and fast," she says, "but I wasn't dilating at any significant rate. So they just kept hanging the drip, hanging the drip . . . It all just felt like I was rolling a ball up a hill. It never felt like I was doing what my body wanted to do. It felt like a fight. Like a fight the entire time."

Although the long, laborious induction did eventually result in the birth of her son, now six years old, Alex is still visibly upset as she tells me how the experience shook her faith in her body and its ability to perform this most "natural" of tasks. "I felt hugely disappointed," she says. "I had so much anger, and I think I still hold on to it. I felt really let down by my body, and I felt out of control, for someone who's quite used to being in control. I was really looking forward to labor, I kind of tackled it in the same way I did marathon training—like, it's going to be hard, but I'm going to get through it. And I felt really privileged that I was going to get the chance to do it. In no circumstance did I think that my body wasn't going to play ball . . . I was most angry at my body."

In contrast, Alex's second labor two years later was spontaneous and breathtakingly quick, with contractions progressing so quickly as she prepared for the journey to the hospital that she ended up deliver-ing her daughter on the kitchen floor among a ragtag group of hastily mustered midwives and paramedics. Recalling the experience, Alex describes a connectedness with her body that was lacking in her first birth: an intuitive feeling that "there were other hormones in the mix. I felt much more *in it* . . . more conscious, much more focused. It felt like whatever pain I was going through, my body was going, 'Here, have some of this as well, because you'll need *that* to counteract *that*.' Rather than the other birth, where it felt like someone was taking jump leads to an engine and you're trying to get it to start and it's not, this one felt like the engine was running and all the fluids were going in the right places. I just knew what to do."

Alex's powerful recollection contrasts the industrial, mechanistic model of the birthing body as a machine to be "jump-started" by artificial hormones with a more primal feeling of deeply held intuition and instinct. Her feeling that her body was offering her "some of this"—a kind of empowering, energizing biochemical antidote to the pain of contractions—is, in fact, supported by strong scientific evidence that the body releases its own natural endorphins in response to unmedicated, spontaneous labor. The brain compensates for the actions of the uterus, which in turn initiates a biofeedback loop that prompts more feel-good endorphins to be produced, which enable the mother to cope with intensifying contractions of the uterus, and so on and so forth until (and even in the moments after) the baby is born. In contrast, synthetic oxytocin does not cross the blood/brain barrier, so while it may have the desired effect on the uterus, it does not activate the brain's reward and pleasure centers in the way that naturally produced oxytocin does. Induction of labor may turn the cogs of the birthing machine—nudging the wary womb into action—but it does not grease the gears with natural endorphins.

While synthetic oxytocin may not have as many immediately alarming side effects as old-fashioned, ergot-based "birth powder," retired midwives Monica Tolofari and Linn Shepherd have spent the last few years raising the alarm about some of the other unintended consequences of these modern "Big Boy Drugs."[38] In a career spanning over thirty-five years in the NHS, working her way up from auxiliary to Consultant Midwife in Public Health and Commissioning, Monica noticed that increasing numbers of inductions using highly concentrated doses of Syntocinon appeared to correlate with a rise in postpartum hemorrhage. She is speaking to me from her home near Birmingham; family photos in neatly arranged frames smile at me over her shoulders, and as we chat, Monica's warmth and pride are tinged with sadness for the battles that characterized the twilight years of her career.

"I went back to work in 2014 in the trust that I actually trained in," she says, "and I found that we were just normalizing [postnatal] blood loss. You know, years ago, if you had blood loss of maybe one thousand milliliters, you'd really, really, really be upset. And then when I returned to that trust, losses of around three thousand milliliters were just being normalized." To put this into context, the average circulating blood volume of an adult female is roughly 5,000 milliliters (5 liters), so to normalize losing 60 percent of the blood in your body after childbirth is a stretch of even the most gruesome imagination. Such a massive hemorrhage can cause severe anemia, with symptoms including dizziness, palpitations, and fatigue that may be debilitating enough to require transfusion. These issues can make physical and emotional recovery even more challenging and prolonged than usual. Uncontrolled postpartum hemorrhage can necessitate extreme measures such as hysterectomy, and in some tragic cases, the bleeding cannot be stopped; the World Health Organization, America's Centers for Disease Control, and the UK's MBRRACE reports into maternal mortality consistently list hemorrhage as one of the leading causes of maternal death.

Determined to find out why the hemorrhage rate had skyrocketed in her trust—and sure that these bleeds were not simply caused by a bizarre local cluster of dysfunctional wombs—Monica sought the help of Dr. Gareth Leng, professor of experimental physiology at the University of Edinburgh.

"I spoke to Professor Leng," Monica recalls, "and he told me that if you give too much [synthetic] oxytocin, you shut off the receptors in the uterus. That will explain the bleed. So the more you give, the worse it becomes." In other words, while a synthetically induced uterus may contract enough to initiate and sustain labor, in some cases it may eventually stop receiving these pharmacological messages to contract. The muscular body of the organ thus becomes lax or "atonic" and, in this slackened state, is more likely to bleed during the final interval between birth of the baby and delivery of the placenta and its membranes.

Monica also attributes this increased rate of hemorrhage to three key characteristics of the modern labor ward: a desire to move women through the overburdened, under-resourced system as quickly as possible; the adoption of electronic, pump-driven IV "drips" that release hormones in a more continuous stream than the tiny pulsations normally produced by a spontaneously laboring body; and regimes that often use higher doses and concentrations of oxytocin than those licensed by the drug's manufacturer.

"It's like flogging a [dead] horse," Monica says of these powerful, pump-driven infusions. "Eventually, the woman bleeds, and all in the name of getting put through the labor ward."

Realizing that there was a problem, Monica began to wonder if midwives at other hospitals had observed these effects of potent, off-license oxytocin regimes.

"I started to look at other trusts to think, well, maybe we're just an outlier," Monica recalls. "Maybe we're the only ones. And I did a Freedom of Information request and found that not all [trusts], but most—something like ninety percent—are following the [higher] guideline. And I was thinking, what? When did that start?"

Monica then enlisted the help of her friend and former colleague Linn, whose own research and FOI requests found that the postpartum hemorrhage rates in some British trusts were as high as 50 percent of all births—exponentially higher than the rates of 3 to 5 percent that were prevalent when she and Monica began their careers in the eighties. Joining our call, Linn's voice lilts down the line in an accent that belies her Scottish upbringing and early training in Glasgow's notorious Rottenrow Maternity Hospital, but her soft tones are seamed with anger as she tells me how she's witnessed a rise in what she perceives as misuse of synthetic oxytocin over the course of her career.

"There are good reasons for choosing the licensed dilution and the licensed dosage range—certain cephalopelvic disproportions [when the baby's head is not a good fit with the mother's pelvis], your sluggish labors—you can improve contractions that have eased off where you don't have any other reason," Linn explains. "And the license was given on the back of these improvements in obstetric care. And then the doctors starting using it like sweeties: like, 'Oh, have another one. Oh, let's double this.'" Reflecting on the steeply rising trend for higher and higher hormone doses in consultant-led labor wards across Britain and Ireland, Linn says, "They just thought, 'Let's go for it.'"

Monica and Linn's outrage at these newer, higher-dose regimes was borne not just from a sense of clinical duty but from a dawning awareness of the long-lasting trauma experienced by women who had bled excessively after induced labor and birth.

When Monica spent some time working in community clinics, "I heard the stories of women who'd had hemorrhages, and it joined the jigsaw. I was shocked how it affected the women, and for how long. I hadn't really thought about the impact on breastfeeding, the impact on the woman, the impact on the next pregnancy, the impact on the family, and on the sisters who wouldn't have children yet but were wanting to. It was a multiple thing, and it was really, really shocking."

Thus began a campaign to encourage hospitals to abandon their patchwork of widely varying induction protocols and to return to the gentler dosage licensed by the drug's manufacturer.

Monica went to her own trust, but, she says, "They just shut me down . . . I tried to [provide] individual care, but the establishment came down on me like a ton of bricks, because you're to follow guidelines exactly as they are."

She and Linn then took their concerns to the Medicines and Healthcare Regulatory Agency, the Royal College of Obstetricians and Gynaecologists, and the Royal College of Midwives.

"We exhausted all the options," says Monica, "but instead of involving us and being proactive, they . . . changed guidelines to accommodate what they're already doing." She admits that the response has been slightly better in America, where dangerously high regimes have become commonplace in a highly pressurized but litigious maternity care system. "What is happening in America is that the lawyers are coming aboard," Monica says. "They are ahead of us in terms of looking at the regimes and actually asking women if they want to sue."

While Monica and Linn accept that the battle to reintroduce gentler, licensed dosages of synthetic oxytocin may be long, laborious, and ultimately unsuccessful, they are both keen to emphasize their dearest wish: that women and birthing people should at least be able to make an informed choice about being induced with unlicensed hormone regimes. Data suggest that roughly 80 percent of women in the UK and US become mothers by the end of their childbearing years;[39,40] if approximately 30 percent of these women have their labor induced or augmented by synthetic oxytocin, with an even higher but as yet undetermined proportion receiving the hormone to expedite delivery of the placenta, then hundreds of thousands of women— and uteruses—are being medicated without a full explanation of that medicine's licensed and unlicensed use, or of the respective risks and benefits of each regime. Surely, argue Linn and Monica, if synthetic oxytocin is the most common drug given to alter the action of the uterus, then shouldn't the owner of that uterus be able to make an informed choice to accept or decline an off-license dose of that drug?

"I'm optimistic that if women understood, they would want the licensed dosage range," says Linn. "There's research that says that when you average out the expectation of making labor shorter by using more oxytocin, you only shorten it by two hours. So my argument is that if you don't mess it up yourself by using synthetic oxytocin improperly, the labor might last that extra two hours but you didn't have unnecessary [interventions] or postpartum hemorrhage. You've gained your obstetric future almost intact. You've gained your postpartum period with as little likelihood of complication as you could possibly manage." She is optimistic, in spite of the many obstacles to her campaign thus far, that birthing people will ultimately prevail once they learn the full extent of oxytocin's misuse and its consequences. "I do think change will come from the women in the end," she says, "because that's where the power lies."

Successful campaigns to limit or eradicate other potentially harmful gynecological practices show that grassroots change is indeed possible, albeit often slow and painful. Pelvic mesh repairs, initially embraced by many practitioners as an effective treatment for certain kinds of uterine prolapse and urinary incontinence, are now far more restricted following vociferous activism by those who suffered debilitating injuries as a consequence of such procedures. Recently, a UK-based campaign has highlighted the need for pain relief (or at the very least, the offer of it) during hysteroscopies—invasive investigations which some people tolerate well without medication but others find incredibly painful and traumatic.[41] Like prolapse repairs and hysteroscopies, it seems that the use of synthetic oxytocin has become a widely accepted part of clinical practice that may now—prompted by the voices of patients and practitioners alike—warrant much closer scrutiny. It may be time for an attitude of cautious reevaluation to replace the gung-ho, "Let's go for it" enthusiasm described by Linn Shepherd.

As we make the requisite small talk that seems to conclude every video call in our pandemic age, Linn fires one last parting shot. She admits that, out of a desire to protect her friend's career, she has thus far suppressed an impulse to wage a high-profile media campaign exposing oxytocin misuse.

"Because Monica was employed by the NHS [at the time], I held off, but I would have gone straight to radio. And I am still ready. I've been the one rattling cages."

To paraphrase the old adage that hell hath no fury like a woman scorned, I would argue that few people possess the fury, determination, and deeply embodied wisdom of a retired midwife with an axe to grind. If you happen to hear a thistly Scottish lilt on the radio while you're reading this book, it might well be Linn deciding that now is the moment to rattle her cage, state her case, and put things right for the mothers, the fathers, the families—and the wombs.

—

IN THE MEANTIME, THE SEARCH FOR THE GOLDILOCKS contraction—that elusive wave of muscle fibers firing in perfect synchrony in exactly the right way at exactly the right time—continues, and researchers worldwide keep seeking a magic elixir to "solve" the perpetual problem of the uterus and its unpredictable ways. A millennium may have passed since women discovered the uterotonic properties of ergot, dispensing it in carefully measured grains, tinctures, and thimblefuls, but its modern and ostensibly more sophisticated equivalent of pump-dispensed, man-made hormones still carries significant risks. A review of these issues by medics from Tennessee's Vanderbilt University in 2020 suggests that side effects like high blood pressure and irregular heart rhythms are increasingly unacceptable in a pregnant demographic that includes rising numbers of women with pre-existing conditions of the heart, lungs, and circulatory system.[42] For that reason, say the study's authors, science should focus its attentions on developing the next generation of "birth powders" and "Big Boy

Drugs": medicines targeted specifically at the cells in the myometrium (muscle wall of the womb). Others suggest that novel treatments might be derived from one of the fifteen plant families known to affect contractions of the uterus; these may not only yield safer alternatives to synthetic oxytocin, but their production could be more economically and environmentally sustainable, bringing a new source of income to developing countries rich in these valuable natural resources.[43]

As with so many areas of women's health—so many, in fact, that this mantra should probably be printed at the top of every page in this book—more research is needed. Until newer, safer alternatives to ergometrine and synthetic oxytocin are developed, and until there is a consensus about the safest timing and method of induction, obstetrics continues to stand at the bedside of the laboring womb, scratching its head, willing it to contract a bit less or a bit more, a bit sooner or a bit later, or just a bit *better*. Doctors who bear the legacy of Gooch, Stearns, Dale, and Moir continue to prescribe "Pit" and "Synt," as the synthetic hormone is often called, sometimes to magnificent and lifesaving effect, sometimes with complex and troublesome consequences. Midwives like me continue to snap those glass ampoules, run those pumps, and gaze anxiously at the clock on the labor ward wall, one hand resting on the swell and rise of a tightening abdomen, one hand poised to pull the emergency buzzer and summon the blue-scrubbed "team."

Perhaps, though, in our rush to prescribe, titrate, and now re-evaluate synthetic oxytocin, we are missing the point entirely. Perhaps it is easier to scapegoat the uterus, and to see it as a problem to be managed and manipulated, than to examine the many other factors that can influence the onset and progress of labor. As Rachel Reed writes, in today's industrialized maternity system, "Any complications are considered to be caused by the woman's body malfunctioning rather than her environment and interventions."[44] It may be easier to posit the uterus—that stubborn, tricksy, mischievous muscle—as the enemy of its own success, rather than to query the effect of the system in which it is expected to perform.

Oxytocin is often referred to as the "shy" hormone, and for good reason: the body produces it most freely, whether for orgasm or for birth, in circumstances that feel safe, intimate, and private. Unfortunately, the typical labor ward (and the preceding trip to the hospital) seldom facilitates such feelings: the uterus is expected to perform admirably through traffic, speed bumps, and car parks, then in a busy assessment area, and then in a room that might be filled with clinical equipment, bright lights, and the searching eyes and hands of strangers. It would be overly simplistic to suggest that every milliliter of Syntocinon or Pitocin could be spared if only the birth space was warm and fuzzy, but it would be equally naive to pretend that the environment and its caregivers have no effect on uterine contractility. The unmistakable tang of hospital bleach and blood; the rustle of paper curtains; the cold steel of speculums and stirrups; the invisible but no less palpable pressure of institutional momentum—all can conspire to make a womb, a woman, and yes, even a midwife, more than a little anxious.

FORTUNATELY, FOR EVERY MIDWIFE CLUTCHING HER AM-poule of oxytocin, there is a woman whose body possesses the wisdom of the many mothers who came before her, passing through the fire that is now hers to endure. The uterus—at the best of times—does what it has evolved over millennia to do, tiny bolts of lightning jumping from cell to cell until the mighty muscle bears down to bring a baby, slick and squealing, into the world.

Loss

A MOMENT OF STILLNESS

*I cried a lot, but it's over, there is nothing else that can be
done except to bear it.*

—Frida Kahlo, letter to her doctor
after a miscarriage[1]

J ust a moment. A pause, if you will, before we move on. Not the
standard minute of silence observed publicly for tragedies that
rock a nation: the flag at half-mast; the murmurs that ripple across
an office or a shopping center before the showy bowing of heads; the
studied solemnity of a newsreader's face. No; a moment, please, for
the private sorrow. The personal tragedy. The indescribable loss of a
baby. This is a silence made all the more painful in its contrast to the
noise that should have rung loudly round the birth space: a raucous
first breath, a mother's cry of exaltation and relief.

Sometimes the womb does wrong. I am (as you may have guessed)
an advocate of this organ and a celebrant of all it can do, but I have
seen it falter more times than I would like to recall. It would be disingenuous to pretend otherwise. It's OK, though—even essential—to

ask why the worst happens. The questioning is part of the grief, and the answers—and the language we use to frame them—are part of the healing.

—

SOPHIE MARTIN WAS ON THE BUS TO WORK WHEN SHE was told her body was incompetent. Eleven weeks prior, she had already been through one of the most physically and emotionally devastating events a person can experience: the loss of her much-longed-for and already much-loved twin boys, Cecil and Wilfred, only twenty-one weeks and one day into their gestation.

It had all started, as episodes both fatal and benign often do, with some light bleeding. Sophie—a midwife herself, and well versed in the pathways of pregnancy care—went to the hospital for what she assumed would be an uncomplicated examination. By this point, she was accustomed to a regular schedule of intimate and often invasive investigations; the twins were conceived by IVF, a process whose success requires almost constant surveillance. Instead of another all-clear, though, the cold pressure of the doctor's speculum was followed by the shock of unexpected findings: without warning, her cervix had already begun to dilate. There had been no contractions—Goldilocks-perfect or otherwise—to let Sophie know what was happening. In the hours that followed, she became hospital property: tagged, cannulated, bled, and monitored, and within days, the pains began.

The inconceivable—but at this point, inevitable—soon happened. Sophie and her husband spent hours holding and loving their boys, who were born too early to survive for more than a short time. The Martins left the hospital with cards bearing their sons' inky footprints, photos of Cecil and Wilfred wrapped side by side in fluffy white blankets, and empty arms.

Nearly three months later, with the emotional pain of bereavement made worse by the physical pain of an infection caused by retained placental tissue, Sophie asked her medical team for a hysteroscopy—a procedure in which a thin telescopic device is passed into the womb—to check for any permanent scarring. Fortunately, there was no sign of long-term damage, but there were other surprising findings: even in its nonpregnant state, Sophie's cervix—the thick, fleshy tube at the bottom of the uterus that's supposed to stay long and closed—was unusually short. As a midwife herself, Sophie knew immediately that this anomaly must have played a part—if not a leading role—in the extremely preterm birth of her twins.

"I was on the bus from my clinic," she recalls during our conversation almost three years after her loss. "I emailed a consultant where I work and was like, 'Nick, my cervix is only two centimeters [long].' And he was like, 'Yeah'—he knew what had happened—'you've got an incompetent cervix.'"

In most cases of miscarriage, stillbirth, or early infant loss, the uterus is not to blame. In fact, the uterus is so rarely to blame—perhaps only once in every hundred such losses, although the figure is hard to define—that it is tempting to omit these events from a celebratory paean to the womb and its capabilities. The leading causes of loss include chromosomal abnormality of the fetus, maternal infection, clotting disorders, and complications of medical conditions such as high blood pressure or diabetes.[2] Sometimes, though, the womb—or, to be more specific, its "neck," the cervix, which adjoins the vagina—causes the loss by dilating painlessly before the fetus has reached a viable gestation at or around twenty-four weeks of pregnancy. As silently as the deep, black well of a pupil dilates to seek light in darkness, the cervix in these instances softens and slips open, just a centimeter or two to begin with, then more, and more again. It may not surprise you to learn that this phenomenon has been given a name that pours scorn on the organ—and the woman—it describes: "incompetent cervix."

The exact origin of the term "incompetent cervix" is unknown, but the condition was first described by seventeenth-century physician Lazarus Riverius as a state wherein "the orifice of the womb is so slack that it cannot rightly contract itself to keep in the seed."[3] This phenomenon was repeatedly observed by obstetricians (and, no doubt, by midwives) in the centuries to come, and at the time of writing, it is defined by the Royal College of Obstetricians and Gynaecologists as "painless dilatation and shortening of the cervix in the second trimester of pregnancy, resulting in pregnancy loss or delivery."[4] The RCOG points out that this diagnosis is often made in retrospect, once other possible causes for the loss have been investigated and excluded, although there are some risk factors that are now known to predispose women to cervical incompetence (or, as it is now sometimes called with a slightly milder but still accusatory tone, cervical insufficiency). Certain connective tissue disorders such as hypermobility or Ehlers-Danlos syndrome can affect collagen levels and elasticity in the cervix. Previous surgical treatment can increase risk, too: people who have had cervical biopsies, laser excision of cancerous or precancerous cervical cells, or Cesarean sections during labor are thought to be at higher risk of painless preterm dilatation.[5] Results of one study of over thirty-four thousand women in California suggest that race may also be a factor, with Black women more than three times more likely than their white counterparts to experience cervical insufficiency, although the reasons for this difference—as with so many racial disparities in healthcare—are still under-researched and poorly understood.[6]

For any woman, and for Sophie Martin, the label of incompetent cervix is a distressing bolt from the blue. Not only is the diagnosis completely unexpected, but the pejorative connotations of the term add insult to the worst kind of injury, implying that the birthing body itself is inherently flawed ("incompetent") or not enough ("insufficient"). Even the occasionally used alternative term "weak cervix" connotes a lack of strength or resolve, a deeply wounding notion to those women who would have done anything to hold fast to their pregnancies. Tommy's, a UK-based charity dedicated to researching and preventing baby loss, says on its website, "Women have said to us that they suffer from feelings of guilt and self-hatred when they go through a late miscarriage or premature birth caused by cervical incompetence." Continuing with crashing understatement, the site states that "some women do not like the term 'incompetent cervix,' but it is used as a medical term. It does not describe you or your body."[7] Herein lies the problem: the term "incompetent cervix" *does* describe a woman's body and, by extension, the woman herself. To pretend that it does not is to perpetuate a false comfort, and to minimize the very real pain caused to women who have been labeled in this way.

When I ask Sophie Martin how she felt when she received her diagnosis during that fateful bus journey, she is—at least at first—ambivalent.

"In a way," she recalls, "I was relieved. You're like, oh, thank God, I know what the problem is now." With her midwife hat firmly in place, she explains to me how good it felt to have an explanation for the inexplicable, and to have a problem that, in her words, "we could fix." However, as our conversation continues, her composure slips and the raw vulnerability of a grieving mother shows through. I ask Sophie whether the term "incompetent cervix"—that is, the language itself—contributed to any lingering feelings of guilt or shame after the loss of her sons. Her answer, then, is unequivocal.

"Massively," she says. "I feel like my body's let me down at every step of the way. I couldn't get pregnant, and I couldn't stay pregnant, and then I had retained placenta for eleven weeks. It's like, I couldn't do a single thing right. I felt really upset and angry that I couldn't just have one thing go wrong—I had to have *everything* go wrong."

This crisis of confidence continued to plague Sophie when she conceived again during a subsequent cycle of IVF. Having had no warning signs that her cervix had been dilating even before the onset of contractions, the prospect of labor in her second pregnancy—especially the prospect of unexpected preterm labor—felt like an axe waiting to fall.

"In the whole pregnancy," explains Sophie, "you're just waiting to go into labor. It was terrifying. As a midwife, I always think labor is so amazing and so positive, but from my personal experience, labor means that someone's going to die. You know, I went into labor, and my babies died." Having her body described as incompetent—or even the marginally less offensive "insufficient" or "weak"—only exacerbated her fear of a womb that could betray her and her unborn child at any time.

Some women with a history of painless preterm dilatation are offered treatment such as pessaries (ringlike devices used to maintain closure of the cervix), vaginal progesterone supplements, or both. Bed rest—an old-fashioned but arguably intuitive way of preventing preterm birth—has not been found to be effective, and as such, is less widely recommended than perhaps it has been in the past. In Sophie's case, she felt that a surgical procedure called transabdominal cerclage, or TAC, would give her the best possible chance at a better outcome: birth of a live baby at a viable gestation. This operation, often performed in the late first or early second trimester, involves an incision through the abdomen and the placement of a strong band around the cervix, ostensibly to hold it closed and prevent any dilatation whatsoever. Because of this closure, the fetus must later be delivered by Cesarean section. Other forms of cerclage—also known as a "rescue stitch" or "rescue suture"—can be placed via the vagina and removed at a later stage to allow for vaginal birth. While not an absolute guarantee against preterm labor, such interventions are generally described as having an 80 to 90 percent chance of success.

Although she was grateful for the opportunity to have cerclage in her second pregnancy, Sophie is understandably exasperated that her twins' deaths could have been prevented had her cervical length been checked and managed earlier in her first pregnancy.

"I felt like it was such a waste," she says. "Like, I could have had a stitch, and this never would have happened."

What she's heard from many other women who have borne the "incompetent cervix" label is that such measurements are seldom done, and even when a short cervix is detected—either incidentally during a routine ultrasound, or otherwise—some doctors adopt a hands-off approach that she sees as dangerously cavalier.

"I really get cross about the treatment of incompetent cervix," she says, "because so many doctors just want to do the 'watch and wait.' They just want to measure your cervical length [again], or say, 'Have a bit of progesterone.' That might work for some people, but there is a chance that those babies won't come home. The amount of women whose doctors say, 'Just watch and wait,' and then they go on to have two or three second-trimester losses . . . I just think, what a waste." For Sophie, this sense of injustice, and of easily preventable tragedy, characterizes her very personal battle. "I went for a checkup with another doctor and she said, 'Well, why did you have the TAC? You only lost one pregnancy.' And I just turned around and said, 'Oh, is two dead babies not enough?'"

Dr. Katie Morris acknowledges that a diagnosis of cervical insufficiency can lead to "lots of uncertainty and worry," but argues that even with the benefit of a growing body of research around the condition, the development of a clear and effective management pathway remains challenging. "Due to the complex multifactorial process of second trimester miscarriage," she says, "we have yet to establish a reliable test of who will benefit from each of the interventions." This "test"—a way of identifying and treating women at risk—is Katie's current goal. As professor of obstetrics and maternal fetal medicine at the University of Birmingham and an honorary consultant in that specialty at Birmingham Women's and Children's Hospital, Katie is leading the C-STICH2 study, a randomized controlled trial of rescue cerclage as a means to prevent miscarriage and preterm birth.[8]

Currently, she says, "It can be difficult to get accurate information on the recurrence risk and we have limited preconception counseling pathways to support women."[9] Hopefully, by the time the eight-year trial is completed and analyzed, there will be a clearer and more compassionate alternative to the "watch-and-wait" strategy that Sophie and her peers have found so dangerously inadequate.

In the meantime, stories like Sophie's illustrate the ways in which—in the small proportion of losses caused by the mechanisms of the uterus itself—current management fails to address the complexity of birthing bodies and the emotional needs of parents who may be grieving a previous loss and continue to live in fear of another bereavement. Would it be easy to check cervical length as part of the other investigations that are routinely performed in pregnancy, and could this identify some women at risk while providing reassurance to others? Could healthcare providers give more weight to women's feelings around the risk of a subsequent loss, and could they proactively manage those risks? Undoubtedly, the answer to those questions is yes, yes, and yes.

Perhaps the easiest question to answer, though, is whether we can use less negative, judgmental language to describe an involuntary physiological event. It hardly takes a master linguist to come up with more appropriate and less injurious terminology. In this chapter, I have chosen at times to use the term "painless preterm dilatation," for which one could even use the easy acronym PPD (and God only knows how medicine loves an acronym). This language is descriptive but neutral. It does not blame the body, nor does it attribute blame, nor add insult to the most painful of injuries. Adoption of this terminology or similar can be done immediately, globally, and at no cost, but with potentially enormous benefit to maternal mental health.

As for Sophie, enraged as she may rightfully be about the ways in which women's wombs are maligned and mismanaged in their most vulnerable moments, her plea for change is punctuated by the most deliciously sweet gurgles, coos, and squeaks. Just out of shot during our video call is a seven-week-old boy called Percy. He is Sophie's, and he is perfect, and the image of Sophie breastfeeding her son while honoring the memory of his brothers with love, fury, and steely determination is deeply moving. This time around—with judicious intervention and more than a little desperate hope—Sophie's uterus did everything it should and could have done.

--

I HAVE HELD THE HANDS AND DRIED THE TEARS OF TOO many women like Sophie. I have chaperoned what should have been routine examinations, and I have watched doctors shudder at the unexpected sight of a cervix that is open when it should be closed, and I have stood by, my features arranged in studied neutrality while my heart thumped hard against its cage, as those doctors have given the news we both knew was coming:

"The neck of the womb is open."

"Yes, it's too soon."

"No, there is nothing we can do."

"No. I'm so sorry. At this stage, your baby cannot survive."

If that sounds harsh, it's because it is harsh, and it is the way we are taught. There are protocols and classes and online modules about giving bad news, and each one of them exhorts the caregiver to be compassionate but unequivocally clear. There must be no false hope or ambiguity. It is important for the message to hit its mark.

There is another message, though, that often goes unspoken, although it is no less important. Losing a baby at twenty-one or twenty-four or twenty-eight weeks—or at any time at all—does not mean that a womb, or a body, or a person, is incompetent or insufficient or weak. It means that the person is human, and humanity is a terminal condition whose course is notoriously unpredictable and wild.

A moment, then, for the womb that yields before the life inside it is ready, for the babies born into stillness, and for the hurtful words that rush in to fill that void.

A moment. And read on.

Cesarean

THE WOMB AND THE KNIFE

The uterus now came into view, dark and livid.

—ROBERT DYCE, CASE OF CAESAREAN SECTION, 1862

On the morning of April 10, 1888, Catherine Colquhoun stood at the bottom of North Portland Street and contemplated the journey that lay ahead of her. Later generations of Glasgow women would name the steep incline "Induction Brae"; local lore had it that if you weren't in labor at the bottom of the hill, you would be by the time you'd hiked your way to the Glasgow Maternity Hospital that sat, grandiose and stately in her sandstone skirts, at the top. Twenty-seven-year-old Catherine was already in the early stages of her first labor, though, her belly swelling with every few steps along the rain-slicked cobbles. If anything, she might have wished for the pains to ease or stop altogether. At just over four feet tall, and with a pelvis severely narrowed by the rickets that plagued so many of the city's slum-dwellers, Catherine knew that giving birth to the child who still squirmed and rolled inside her body could easily cause her own death. Not for her the home birth that awaited most of Glasgow's women, comforted by familiar sights and smells, delivered by the local "howdie" and cheered by siblings and neighbors. As Catherine finally summited Induction Brae, she paused awhile by the pillared portico of the hospital and took one last look at the city spread out like a grimy shawl at her feet. She was a daughter of Glasgow, but her child would have a birth like no other; instead of hunkering down among family by the warmth of her own hearth, Catherine braced herself for strange voices and curious hands.

Once inside the building, Catherine was no longer just another "wee wummin" with her woolen wrapper drawn tight against the morning drizzle. She was stripped naked, probed and scrutinized; her body, small and twisted as it was, held the promise of innovation and fame for the man who assumed her care. Professor Murdoch Cameron, one of the hospital's chief obstetricians, surveyed Catherine and described her in his notes as "a little woman, somewhat delicate, and with all the appearance of a patient deformed by rickets in a very marked degree."[1] After examining her internally, Cameron determined that the inner diameter of Catherine's pelvis was no more than an inch and a half across, making any attempt at vaginal delivery futile and inevitably fatal for both mother and child. He then called for his colleagues, Drs. Sloan, Reid, Oliphant, and Black, who each performed the same intimate assessment in turn—strange hands in a private space, over and over again—and each reached the same conclusion. Catherine's reaction to such indignity is not recorded, although any person who has felt the searching touch of an unfamiliar practitioner during labor can imagine the paralyzing horror of this fivefold examination. By this time, several hours had passed, and Catherine's pains continued as the medics reached their consensus: Cesarean section was the only reasonable course of action.

At 4.30 p.m., having consented, as Cameron later wrote, to "any operation that might be advisable," Catherine was assisted onto the table of the hospital's operating theater and a rubberized mask placed over her nose and mouth. In the final moment before she was cast into a fitful chloroform sleep, Catherine would have seen Cameron peering down at her in the flickering gaslight, his gaze intent and focused behind his wire-rimmed spectacles; behind the surgeon, the shadowy figures of the other men who had searched her body and declared it perfectly flawed; and just visible at their backs, row upon row of wooden benches filled with students hastily summoned to observe this pioneering procedure.

In fact, Cesarean section had been performed many times before, with descriptions of this kind of birth found in ancient Greek, Egyptian, Hindu, and Chinese texts. While the name of the procedure is commonly thought to reflect a belief that Julius Caesar was delivered abdominally, it is more likely to be a reference to the emperor's decree that all babies should be extracted in that way if the mother's death in childbirth seemed inevitable. For centuries, infection and blood loss made the procedure so dangerous that it was used only as a last resort. Some form of Cesarean section evolved in cultures around the world for exactly these scenarios, with occasional reports of maternal and fetal survival as surgical techniques improved. There are nineteenth-century accounts of Cesarean sections being performed by Indigenous people in Rwanda and Uganda, using botanical painkillers, careful hygiene, and wound closure with locally available materials; and in Cape Town in 1820, British army surgeon James Barry performed a Cesarean section that saved the lives of both mother and child.[2] (Incidentally, Barry was found after his death to be physiologically female; whether his gender journey gave him a particular interest in or sympathy for the plight of women is a matter for speculation.) For the most part, though, surgeons regarded the Cesarean section with wariness borne of years of tragic outcomes. Writing in

1862, Robert Dyce, professor of midwifery at the University of Aberdeen, describes an attempt to deliver a stillborn fetus by Cesarean section. Reflecting on the subsequent death of the patient—a small-statured woman like Catherine Colquhoun—Dyce lamented "another case added to the melancholy list of unsuccessful operations on the parturient female."[3] The outlook continued to be bleak right up until Catherine Colquhoun's arrival at the top of Induction Brae. Change drew nearer, though, with every tightening of Catherine's womb in her narrow, twisted pelvis.

Having trained at the nearby Glasgow Royal Infirmary under Joseph Lister, the pioneer of antiseptic technique, Murdoch Cameron believed that Cesarean birth could now be achieved with minimal risk of the infections that had blighted British hospitals and continued to kill thousands of their patients, from women to the war-wounded, every year. The lives of mothers and babies depended on the death of bacteria that so often festered and multiplied in surgical wounds. Lister's work in general medicine suggested that this battle could be won in obstetrics, too, and Cameron was ready to fire the opening shot.

As luck and science would have it, Catherine's Cesarean went entirely to plan. Cameron opened and closed the abdomen with instruments and "No.3 Chinese twist silk ligatures" soaked in Lister's tried-and-tested carbolic acid solution. In fact, Cameron was so dedicated to surgical antisepsis that when a bottle of ether caught alight in the operating theater, the professor carried on with his work undeterred, insisting that the blaze would simply help to sterilize the environment. The patient, he noted in his report in the *British Medical Journal*, endured her anesthetic "well and without sickness"; blood loss was minimal, thanks to an injection of ergot extract, and a six-pound, twelve-ounce baby boy—aptly named Caesar Cameron Colquhoun—was safely delivered. In the recovery bay, nurses ministered to Catherine's every need, warming her with "hot pans" placed around her body and feeding her teaspoonfuls of iced milk and soda every hour. By the fourth day, she was allowed greater sustenance of "chicken soup, fish, eggs and beef-tea," and her temperature, pulse, bowel motions, and bleeding were charted carefully until her release from the hospital on May 16, by which time young Caesar had gained two pounds. A photograph of Catherine taken upon her discharge shows a young woman with the full breasts and fresh wound of any modern mother in the early stages of postoperative recovery. She is healthy, if hollow-eyed, with lustrous dark hair braided in neat coils around her head.

Catherine Colquhoun may have been small of stature, but her participation in Cameron's surgical experiment—Patient A, as it were, for the antiseptic Cesarean section—stands as a monumental contribution to modern obstetrics. Murdoch Cameron went on to perform at least a dozen such operations, two of which were on "rachitic dwarves," as he called them, with bodies similar to Catherine's, and the rest on women whose labors would have been fatally impeded in some other way. By 1901, Cameron's success gave him the confidence to declare, "I think the time has come when the lives of mother and child alike may be saved":[4] a stark contrast to the history of Cesarean section as a last resort, to be used only in cases of certain impending death for one or both parties. In a city often derided for the filth of its slums and the wretchedness of its impoverished inhabitants, the Glasgow Maternity Hospital became renowned as a pioneering center for clean, safe, surgical birth, and that legacy continues today.

Cesarean sections have become so commonplace that, in the UK and US, approximately one in three women gives birth in that way. In some other countries, the rate is much higher, hovering just over 50 percent of all births in Egypt and Brazil.[5] Globally, the C-section—or "lower uterine segment Cesarean section," to give it its proper medical name—is one of the most frequently performed operations in the world. It may well even be the most common operation, but because many countries often exclude obstetric procedures from their nationally collected surgical data, it is difficult to be certain. Regardless of any statistical discrepancies, one fact is clear: if you live in the developed world, and you own a uterus, and you use that uterus for pregnancy, then there is a pretty decent chance that it will have had a Cesarean section by the end of its childbearing career.

As common as the C-section is in our twenty-first-century, obstetrically led world of birth, the procedure itself has not changed drastically since young Caesar Colquhoun was lifted from his mother's womb. Sterile operating rooms with dazzling lamps and disposable green drapes may have replaced the gaslights and starched linens of Cameron's day, but the drama performed within that theater of birth remains substantially the same.

Once anesthetized—either from the torso down (with spinal or epidural anesthesia) or asleep completely (under general anesthesia)—the patient is carefully positioned on the operating table. The room may hum with quiet urgency as each of the theater's players assumes his or her role: the obstetrician, who will act as surgeon; a more junior doctor, perhaps, to help and to learn; the anesthetist; various theater assistants and auxiliaries; a nurse or midwife to "scrub," or supervise the preparation and passing of instruments; a midwife to assume care of the baby after birth; and possibly pediatric staff, too. Under these many watchful eyes—and often under the tick of a clock, as each minute in surgery must be documented and accounted for—the birthing person is prepared.

Their skin is cleansed with antiseptic solution and an incision is made in the skin of the abdomen: previously, a "classical" vertical cut from the umbilicus (belly button) to the pubic line; more often now, a horizontal slice opening a thin, smiling seam from one side of the lower abdomen to the other. Whoever is "scribing" during this operation—for there must always be someone, usually a midwife, to record each action—will write the standard line, powerful in its simplicity: "Knife to skin."

This incision is then widened by "blunt extension"—a euphemistic description of what is, in fact, a quite primitive and brutal act, in which the surgeon and his or her assistant use their gloved hands to pull the incision's edges out and away from the center, exposing the muscle layer below. "Pull," perhaps, is euphemistic; a firm, steady tug is necessary to tear the tissues heretofore so neatly and perfectly assembled in the pelvis. Patients are often cheerfully assured before surgery that all they will feel is a sensation "like someone rummaging in a handbag, or doing the washing-up in your belly." These stereotypically female metaphors of shopping and housework are glossy understatements of the force that is used at this point in the procedure.

The tempo changes as the surgeon continues: there are a few moments of quieter, more delicate handiwork as the abdominal wall is penetrated with careful snips and cuts, and then again, the brute mechanics of the procedure come into play as a large, spade-like retractor is used to deflect the bladder down and away from the womb. The uterus presents itself, as pink and shiny as a swollen pearl.

With a few more deft slices of the scalpel, the womb and amniotic membranes are opened and there, like the smallest inner kernel of a Russian doll, lies the fetus, still mute and dusky until it is lifted aloft.

Time stands still for a moment as the surgeon holds the child up for mother and partner to see—a glimpse of just-flushed limbs, and perhaps the sound of a tentative, gurgling cry—and then the steady rhythm of surgery resumes.

The umbilical cord is clamped and cut and the baby passed to its attendants for any necessary assistance and an initial examination while the surgeon returns to the uterus and its mute, open mouth. The placenta is extracted by a combination of medication (our old friend, synthetic oxytocin) and traction (a firm, steady pull), any remaining blood is either cauterized at its source or suctioned away, and the uterus—now already contracted down to a fraction of its pregnant size—is closed in layers of whipstitched sutures. Much to my shock as a wide-eyed student midwife, the uterus may even be "exteriorized" for repair; in other words, some surgeons actually lift the organ out of the abdominal cavity so that it rests lightly on the belly—"All the better to see you, my dear"—working away until the womb is as neatly crimped and closed as a big pink pastry, before nestling it once again in its rightful place within the pelvis.

To the novice observer, the uterus can look startlingly vulnerable in these moments of exposure; to me, the sight of a womb, bare and bleeding, under the hot theater lights remains almost incomprehensibly incongruous. A reminder, in spite of all of my professional detachment, that the scarred uterus in my own belly is among the one in three, and that my skin bears the silver seam of my first daughter's birth. Fortunately, there are other tasks to occupy my attention as a midwife at these times: the baby is usually assessed, weighed, wrapped, and returned to its parents while the doctor continues the somewhat more complicated work of closing the abdominal layers that were so quickly interrupted just moments before. The birth itself often happens within minutes of arrival in theater; "putting you back together," a mother is often told, "is the tricky bit."

–

EVERY DAY, THOUSANDS OF WOMEN LIE GAZING AT THE ceiling while their insides are neatly replaced and repaired; every day, these women meet their babies in this prone position, their hearts surging with emotion while their bodies are numb below the waist. Few of these modern mothers end up in theater for the same reasons as Catherine Colquhoun; most birthing people today are of larger stature than their ancestors, true cases of cephalopelvic disproportion (impossible mismatch between the size of a mother's pelvis and her baby's head) are rare, and rickets is rarer still. However, just as reasons for induction of labor have multiplied since the procedure became possible on an industrialized scale, so indications for birth by Cesarean section have become more varied and numerous, too, since the age of Cameron and Colquhoun. Nowadays, emergency Cesareans (carried out when there is imminent danger to mother and/or child) are done for all manner of distress or delay in labor, and elective (planned) Cesareans are recommended for many reasons ranging from the imperative (a low-lying placenta that might bleed catastrophically in labor) to the more subjective (maternal preference after a previous traumatic labor). Time has also shown that one of the most common reasons for Cesarean section is, in fact, Cesarean section: a woman whose baby was delivered surgically has a statistically higher chance of a similar outcome in her next birth. Cut begets cut, and as Cesarean rates rise around the world, further increases become, to a certain extent, a self-perpetuating cycle. Hence the skyrocketing proportions of women having a "belly birth," in spite of the World Health Organization's recommendation that rates should be kept as low as safely possible. "Cesarean sections are effective in saving maternal and infant lives," advises the WHO's 2015 statement on the subject, "but only when they are required for medically indicated reasons. At population level, Cesarean section rates higher than 10% are not associated with reductions in maternal and newborn mortality rates."[*]

And this, dear reader, is where I would like you to imagine the sound of a needle scratching discordantly off a record. You read it correctly: the world's foremost authority on health just told you that there's no evidence of any benefit to mothers or children—*any* benefit at all—if Cesarean sections account for more than one in ten births. This statement is especially provocative in light of the fact that Cesarean sections account for roughly 30 percent of all births—three in ten—in so many of the world's nations, from the developed to the developing world, from the UK, the US, Germany, and China, to Venezuela, Vietnam, Thailand, and Tunisia. For some demographic groups, the proportion is even higher: among people over forty giving birth in England, for example, 49 percent of all deliveries from 2020 to 2021 were Cesareans.[7] The reality is stark: if you are a member of this generally healthy population, you have a roughly half-and-half chance of giving birth vaginally.

Can so many women's bodies really be so flawed, or so many pregnancies be so severely endangered, that surgical delivery has become the safest option? Why the vast discrepancy between recommendation and reality? All those other operations, the time and money spent in theater, the women with scars like mine, the uteruses opened and shut like so many rummaged handbags—for what? There must, one would expect, be a very, very good reason for these many millions of operations. Either that, or the World Health Organization is wrong.

As one might also expect, the media often suggests its own time-worn reason for this spike in apparently unnecessary Cesareans: the women—not the WHO—are wrong. They are wrong, and they are petty, and they are demanding—according to the headlines— needless surgery for their own nonsensical whims. At the turn of the millennium, the phrase "too posh to push" evolved in the British media in response to the rise in maternal requests for elective Cesar-eans, and women were variously blamed for being too scared of labor, too quick to throw money at private maternity hospitals, too keen to emulate celebrities, and too eager to preserve the integrity of their va-gina or the shape of their child's head (or all of the above).[8] Research subsequently proved that women's reasons for requesting operative birth were far more complex and legitimate than was often suggested by the media,[9] and national US and UK guidelines have since been changed to advise caregivers that maternal requests for Cesarean should be honored if the mother chooses to continue after a full and frank discussion of all pertinent risks and benefits.[10,11] Recently, hospitals in England have been advised to abandon altogether the use of Cesarean section rates as "a metric for maternity services."[12] Some might see this guidance as evidence of the dangerous normalization of intervention, while others may see it as a reflection of increasingly nuanced, productive debate around the issue of operative birth.

Now that the matter appears to have been settled, or at least given the stamp of professional approval, in Britain and America, the media spotlight has turned toward Brazil, the global epicenter of maternally requested Cesareans. The overall C-section rate in that country rose from 40 percent to 55 percent between 1996 and 2011, with some sources estimating that the figure may be as high as 84 percent in Brazil's many private hospitals.[13] A flurry of articles about Brazilian birth culture began to appear in the international press, but instead of focusing on the many impoverished women struggling to access safe healthcare in the country's favelas, the media focused on wealthy women in private hospitals where Cesarean births had evolved into elaborate social celebrations. According to one piece, the São Luiz hospital in São Paulo had begun setting the stage for sections with pre-op banquet rooms full of crystal-vased roses and chocolates on silver trays. Expectant mothers could pay to get their hair and makeup done in the final pre-theater moments, and friends and family could view the operation itself from an adjoining suite with its own balcony and minibar.[14]

The reader of the article is invited to ogle at these vain social-ites and their kin; the phrase "too posh to push" is not explicitly mentioned, but its echo lingers between the lines. In counterpoint, though, one study in which over a thousand Brazilian women were interviewed about their birth preferences suggests that the rise in operative birth is driven less by the petty whims of moneyed mothers and more by very real and legitimate fears of unnecessary interventions in labor.[15] In an interview in the *Atlantic*, Simone Diniz, an associate professor of maternal and child health at the University of São Paulo, describes a macho, misogynistic system in which overuse of induction, episiotomy, and electronic fetal monitoring is often compounded by verbal abuse from uncaring staff: "There's the idea that the experience of childbirth should be humiliating . . . When women are in labor, some doctors say, 'When you were doing it, you didn't complain, but now that you're here, you cry.'"[16]

Such behavior may seem unimaginably cruel, but unfortunately, numerous studies suggest that this kind of abuse may not be unusual, or even exclusive to Brazil. The 2019 Giving Voice to Mothers survey found that of 2,138 American women, 28.1 percent of those who gave birth in a hospital were subject to one or more types of mistreatment "such as: loss of autonomy; being shouted at, scolded, or threatened; and being ignored; refused; or receiving no response to requests for help."[17] Women of color, women with a Black partner, and "those with social, economic or health challenges" experienced consistently higher rates of mistreatment: a depressing demonstration of attitudes toward marginalized and vulnerable people in one of the world's wealthiest and most "advanced" nations. This kind of systematic harm has been identified by clinicians and academics as part of a wider phenomenon of "obstetric violence": a term first coined by Venezuelan researchers in 2010 to describe healthcare providers' dehumanization, abuse, and pathologization of women during pregnancy and birth.[18] Although this behavior is driven by a number of complex factors, including patriarchal, racist, and classist attitudes, and institutionalized standards of care that prioritize the needs of the system over those of the birthing person, the phenomenon itself is all-pervasive: varying forms of obstetric violence have been found to exist in every part of the world.[19,20] In this context, perhaps it should come as no surprise that some women—and, indeed, an especially high proportion of Brazilian women—would see elective Cesarean section as a means of avoiding a potentially traumatic and highly medicalized trial of labor—a choice perhaps not anticipated or intended by the early pioneers of Cesarean section, but a valid choice nonetheless, and not as frivolous or unreasonable as some corners of the media would have you believe. Those theater-side viewing galleries and on-call hairdressers are the glamorous accoutrements of an ugly truth: that we live in a world where major abdominal surgery is more appealing to some would-be mothers than the perils of the labor ward.

AS CESAREAN RATES CONTINUE TO RISE WELL BEYOND the WHO's recommended rate of 10 percent, so women and their caregivers around the world continue to search for new ways of making the procedure safer, and possibly even enjoyable and empowering. For Dr. Ihab Abbasi, a consultant obstetrician and gynecologist in Swansea, the decision to try a new and potentially controversial style of Cesarean section stemmed from a very simple desire to impress his girlfriend.

"The true story," Ihab tells me during a quiet moment in his hospital office, "is that I met someone who's now my wife. She's a counselor and she was a trained midwife as well before that. She deals with birth trauma and she's had a section herself. And we were talking about this thing called 'gentle Cesarean' and she said, 'Why haven't you tried it?' And I thought it was ridiculous, but I decided to give it a go. This was in January 2018, and I'd already been doing Cesareans for ten years, so I went into theater that day thinking that this was the most ridiculous thing I would ever do. I went out of theater thinking it was the most amazing thing I'd ever done. What had I been missing for ten years? The mum was crying, the dad was crying, the staff were crying. I knew it was the way forward. And I've not done any other kind of Cesarean since."

It may sound as if Ihab cast some kind of magical medical spell over the operating theater that day, but the changes he made to his practice were actually very simple.

"It's not a new surgical technique or anything fancy," he explains. "It's just a change in mindset toward making the birth more woman-centered, and putting the spotlight on the mum, rather than on the surgeon, and changing it from an operation to an actual birth experience, just like we do with vaginal birth."

On a practical level, gentle Cesarean—or natural or woman-centered Cesarean, as it's variously called—is a bundle of simple adjustments first developed at London's Queen Charlotte Hospital by obstetrician Dr. Nicholas Fisk, anesthetist Dr. Felicity Plaat, and midwife Jenny Smith. In their 2008 report on this new approach, the team describe how the baby is allowed to emerge slowly and gently from its mother's open abdomen, its body eased out gradually as it would be during vaginal birth; the parents are encouraged to watch the birth, with a view unobstructed by the usual surgical drapes; and skin-to-skin contact in the moments after birth is encouraged, with dim lights and soft music creating a soothing mood and ECG leads, blood pressure cuffs, and intravenous cannulas placed as unobtrusively as possible.[21]

Ihab, who now offers gentle Cesareans as standard to every woman on his elective list, emphasizes that the practice is not dogmatic. Rather, its flexibility is part of its appeal.

"I just change it according to what the woman wants," he says. "Yesterday there was a woman who didn't want any music; fine. And one person didn't want skin-to-skin. That's their birth. It's not a recipe. Every step of it can be designed and adjusted to be more kind and more focused toward the woman's wishes." This facilitation of choice,

Ihab argues, has a powerful effect: "The word 'healing' comes up so many times in the feedback I get. Women talk about how that made them feel part of the birth rather than a lot of experiences they've had in the past, when they'd be lying on their back staring at the ceiling, and hear noises, and then hear a baby crying, and then a few minutes later, they'll see the midwife dressing and wrapping a baby, and then . . . they have a baby."

Nikki Syvret, a midwife and mother of three, is one of those women who opted for a gentle Cesarean after previous traumatic births: in her case, a forceps delivery with a severe tear and a difficult recovery, and a prolonged second labor followed by a Cesarean section performed after much negotiation with a locum doctor whom Nikki describes as "derisory and sneering." Speaking to me from her home in Nottingham while her children dot in and out of the room in search of snacks and entertainment, Nikki recounts the journey to choosing an elective, gentle Cesarean section for her third and final birth.

"I'd been reading a lot of ancillary texts about birth," she says, "and they just gave me a real awareness of how clinical that theater environment is, and how detached it is from a normal birth experience, and how there is so little control in it. There's a complete disconnection. You've got no bodily sensation whatsoever and there's a physical screen in front of you, so even if you can feel anything that's going on, you can only guess what's actually happening. Of course the staff are going to be guided by safety and routine—they're very task-focused—but it should be a magical and spiritual and emotional journey as well."

Having found a consultant who was sympathetic to her desire for a more woman-focused experience, with all the clinical and environmental adjustments that entailed, Nikki ended up having a Cesarean section that was every bit as "magical and spiritual" as she had hoped. Exactly as Fisk, Plaat, and Smith suggested, Nikki's caregivers created a calm, intimate ambiance in theater, and she and her husband watched in awe as their baby emerged slowly and gently from her body. After lifting the baby's head out of the uterine incision, the doctor allowed Nikki's own subtle contractions—brought on by the irritant action of surgery—to ease the rest of the child up and out.

"The incision was made and her head came out," Nikki recalls, "and she was grimacing already and wiggling a bit. Someone moved one of her arms, and once her arm was free, then she started shuffling and rolling, and flexing her legs and pushing. So there was this whole kind of natural emergence with minimal interruption. And because I could see what was happening to me, there was less disconnect. I was seeing what my body was doing with my baby to enable birth to happen. It was very reverential."

As if on cue, one of Nikki's children interrupts our conversation to say that she's tipped over a pile of washing, but even that reminder of the mundanity of everyday motherhood can't diminish Nikki's obvious delight in recalling her third, and arguably most fulfilling, birth experience.

"I felt euphoric," she says.

Back in Swansea, Ihab tells me why this kind of Cesarean is not only kinder for the mother but gentler to the uterus, too. "They hardly bleed," he says of the women who deliver in this way, because of the slower, more controlled birth of the baby from the abdomen. The Queen Charlotte's team describe this as a kind of "tamponade" of the uterine incision, and as Ihab explains, "The body of the baby is blocking the cut, so they won't be bleeding from the edges of the uterus." In other words, just as one might staunch bleeding by applying pressure to a fresh wound, so the baby itself presses against the edges of the Cesarean incision, producing a similar effect. Ihab adds that "the placenta is in place for longer, so there's definitely less bleeding than if you pull the placenta out and close the uterus quickly."

Although some of his colleagues have expressed reluctance to take these extra moments at the table, Ihab argues that this is a false economy of time: "I can't justify not waiting these five minutes. I mean, it's *five minutes*. When another consultant tells me he hasn't got five minutes, I say, 'When you go to theater, do you time it? Do you say, if I'm not out in twenty minutes, call someone else? Or do you go in expecting to take anything between twenty minutes and an hour?' These are just excuses we make because we do not want to change." To foster a more positive attitude to this evolving practice, Ihab says he is spreading the word among the more junior doctors in his department. "I've started with my trainees," he says. "It works better when you plant the seeds."

Although there are few studies comparing the outcomes of gentle Cesareans with operations performed in the traditional way, what little evidence there is suggests that the gentle option (when clinically appropriate, in the absence of emergency) is safe; in some cases, there may even be an improvement in measurable outcomes such as exclusive breastfeeding, postnatal infection rates, and length of recovery.[22,23,24,25] A less formally clinical—but no less important—outcome is the increased maternal satisfaction that appears to be associated with gentle Cesareans.[26] This is not to say that traditional Cesareans are always unsatisfying or unwanted; in fact, despite significant research in this area, there appears to be no clear evidence that mode of delivery has a definitively good or bad effect on postnatal mental health. Increasingly, studies suggest that a birth that feels out of control or mismatched from expectations—regardless of whether that birth is surgical or spontaneous—may be more likely to contribute to conditions such as postnatal depression or posttraumatic stress disorder.[27,28]

I've certainly witnessed a similar phenomenon in my time as a midwife: one woman's "good" birth is another's trauma. What looks straightforward on paper—a quick labor with no interventions, for example—may be unimaginably upsetting for someone who feels overwhelmed by the speed and intensity of her contractions, while a birth that might seem more challenging—a long labor, say, followed by an emergency Cesarean—might feel like a triumph for someone who simply wanted to hold her longed-for baby in her arms. Perhaps, then, for birthing parents who need a Cesarean section, the "gentle" version of the procedure offers a satisfying mix of careful planning and person-centered, compassionate care. When suitable—and when accepted by staff who may be resistant to change—the gentle Cesarean may indeed be the kindest cut for both woman and womb.

Some detractors argue that the reframing of Cesarean as "gentle" or "natural" may be another dangerous step toward normalizing operative birth in an already over-medicalized obstetric system, while others argue that the promotion of this new kind of procedure is both pragmatic, given the seemingly inexorable rise of Cesarean rates, and empowering for the millions of women who may yet choose to deliver in this way.[29,30] Looking back at the conversation around gentle Cesareans ten years after their initial report, Jenny Smith and Dr. Felicity Plaat wrote, "We believe the debate generated by the name to be positive as it demands that we ask why we should not try to optimize the birth experience for women having Cesarean section if putting women at the center of care is the aim."[31]

—

WHETHER A PERSON GIVES BIRTH ABDOMINALLY OR VAGINALLY, IT'S clear that a more holistic approach is needed. The word "disconnect" came up time and again during my conversations with Nikki and Alex. Each woman's birth experience was markedly different from the other's, but both women felt like the modern, obstetric-led environment and the rituals performed within it caused a dissociation from some deeper, primal part of themselves: a part, possibly, that resides literally and metaphorically within the womb itself. Alex described the jarring sensation of having her "engine" jump-started by a relentless stream of synthetic hormones, while Nikki recalled being physically and emotionally numb to that part of her body that existed beyond the draped screen of her first Cesarean section. Each woman's experience is unique, but after observing thousands of births in my career, I can also say that this intuitive desire for connection—and the sense of grief when that connection is lost—is universal. As science continues to strive toward that "perfect" birth, with its obedient womb and its Goldilocks contractions, it is important not to lose our sense of wonder at this most miraculous organ and all it can do in its most powerful moments.

"I have so much appreciation and respect for that organ and the female body, however it does it," says Alex, recalling the birth stories shared by her circle of friends. "When it works, it works incredibly well. The fact that any person who labors can go through that and come out of the other side—and the fact that my friends who've had C-sections can be cut in half and then pick up a baby and go about like it's perfectly normal—it blows my mind."

In her novel *Nightbitch*—a meditation on the struggle to reconcile identity with motherhood—Rachel Yoder riffs on this theme of birth's brutality: "This thing comes from us . . . It rips its way out of us, literally tears us in two, in a wash of great pain and blood and shit and piss. If the child does not enter into the world this way, then it is cut from us with a knife. The child is removed, and our organs are taken out as well, before being sewn back inside. It is perhaps the most violent experience a human can have aside from death itself."[32]

At the heart of this experience—this assault on bodily integrity, this epic struggle, this bloody triumph—is the womb. Whether left undisturbed, coaxed into action, or sliced and sutured, this muscle exerts and endures with little thanks or fanfare. It does its work, and as time rolls from one month to the next, it prepares to start again.

Postpartum

CLOSING THE BONES, TAKING UP SPACE

But what about me? *I whisper*
secretly and to think,
around these parts used to be

the joyful place of sex,
that is now this intimate
terror and squalor.

—BRENDA SHAUGHNESSY, "LIQUID FLESH"

Fatima Abdullah speaks soft words of comfort as she folds seven lengths of fabric around your broken body. You have come through the fire of childbirth, and whether your baby was born alive or still, cut from you or pushed, you need healing, and peace, and nourishment. She has fed you delicacies laced with spices to warm and sustain, and urged cups of ginger tea to your lips. She has massaged sweet oil into the loose skin and fat of your newly empty belly, her fingertips making trails across the stretch-marked terrain of new motherhood. Now, with this ritual wrapping, she honors and restores you. The scratch of green paper surgical drapes, the ward's smell of sour milk and antiseptic, the strange, humiliating paraphernalia of paper pants and peribottles: all fade to black under Fatima's gentle touch.

Bolts of cloth are wound around and up your ankles, legs, hips, arms, and chest, and Fatima pulls just tight enough to make every part of you feel held. "Give thanks," she tells you as you drift to a deeper place. "Thanks to the One that made you. Thanks to the legs that carried you on this journey." You are not sure how long you lie suspended in that liminal space, not so far from the one you passed through during birth itself, but soon you become aware of air on your skin as Fatima unwraps you from the chest down, releasing you back into the present. The last length of fabric is slipped away from your calves, and so the ceremony called by many names in cultures around the world—and now known widely in the West as "Closing the Bones"—is complete.

As an educator and doula serving families in Northern Virginia and the area around Washington, DC, Fatima has supported dozens of people through pregnancy and birth, but in recent years, she identified a need to address the intense vulnerability of the postnatal person, their womb, and their spirit—entities she sees as intimately enmeshed during this transitional phase. The wrapping ritual she performs is integrated into a comprehensive program of postnatal care based on the key principles of heat, bodywork, and support. While Fatima's version of the ritual is based on Al Shedd—a traditional Moroccan ceremony—iterations of this practice have been identified in Indigenous cultures on practically every continent.

Birth may be seen as the main event in the modern industrialized world, but this focus is a relatively new phenomenon. Long before social media and celebrity tabloids began to feature images of "snapbacks"—miraculous weight-loss transformations in the days and weeks after birth—the new mother and her womb were deemed worthy of celebration and care. Postnatal rituals involving some combination of nourishing foods, warming practices, and abdominal wrapping and release have evolved in diverse cultures and locations, from Japan and Vietnam to Malaysia and Moldova.

"These rituals allow the mother to be 'mothered' for a period of time after the birth," writes Cindy-Lee Dennis, professor in nursing and psychiatry at the University of Toronto, in her comprehensive review of such practices. In recent years, this period has been recognized as a formative stage in the life of the birthing person, affecting every aspect of their identity and, as such, deserving of the kind of examination and care that has perhaps been lacking in the industrialized West. Originally coined by anthropologist Dana Raphael, the term "matrescence" now refers to what clinical psychologist Aurélie Athan describes as "a developmental passage where a woman transitions through pre-conception, pregnancy and birth, surrogacy or adoption, to the postnatal period and beyond . . . The scope of the changes encompass multiple domains—bio-psycho-social-political-spiritual—and can be likened to the developmental push of adolescence."[1] Fatima Abdullah suggests that rituals such as Closing the Bones address the enormity of this transition in a way which is unique to each mother: "For many women, [the practice] is about honoring whatever journey they went through. For some people, it's a painful, traumatic journey. And for some people, it's beautiful. Your emotions catch up with what your body went through. It's something that takes time." The fact that similar postnatal rituals can be found from East to West and virtually all points in between is a testament to the universal recognition of the birthing person's needs as they negotiate the crossroads of matrescence.

In addition to marking this transition—painful or triumphant as it may be—postnatal rituals are believed to have important physiological benefits for the body and, more specifically, for the womb. Many cultures identify the postnatal uterus as a site that has become "open" or "cold" during birth, and so must be closed, warmed, or adjusted back into place. Tema Mercado, a Mexican midwife who practices La Cerrada Postparto—a postnatal warming and wrapping ritual remarkably similar to Al Shedd—says, "This process is beneficial to any woman who has experienced a strong opening in her womb area. In Mexico we called this 'Frío en el vientre.'"[2] Thai tradition includes Yu Fai, in which the new mother is wrapped and advised to rest on a bed over a warm fire to aid recovery and uterine healing,[3] while in Trinidad, cloth binders are used to "set" the womb in place and close a passage that was opened and vulnerable during delivery.[4] Layla B. Rashid, a Moroccan birth worker who initiated the revival of Al Shedd, acknowledges that postpartum care is quite literally a matter of life and death: "The elders in Morocco have a saying that the new mother's grave is open for 40 days, because they know how vulnerable she is."[5]

Modern midwives and obstetricians, too, recognize that the postnatal uterus exists in a fine balance of power and peril in the days and weeks after birth. Having fulfilled what some might say is its fundamental purpose—growing and delivering new life—the postnatal womb must now perform a list of complex tasks, each one challenging and potentially risky in its own right. With the completion of the third and final stage of birth—the expulsion of the placenta and its membranes—the uterus must heal a large, open wound site while also protecting against infection, returning to its pre-pregnant size, and regenerating its lining to prepare for another potential fertilization and pregnancy as quickly and safely as possible.

From the outside, the only evidence of this process is lochia—the bloody flow that leaves the uterus via the vagina for approximately four to six weeks after birth. Identical to menstrual effluent in appearance, but markedly different in composition, lochia clears the uterus of amniotic fluid, endometrial tissue, mucus, red and white blood cells, and sometimes even remaining fragments of the placenta and/or amniotic sac. As lochia is expelled, the uterus busies itself with regrowing healthy cells over the exposed placental site—a process which is completed by roughly three weeks postpartum[6]—and begins to involute, or shrink, until it is once again tucked neatly behind the pubic bone. Contrary to popular belief, the rate of involution is highly individual—influenced by factors including age, number of previous children, and mode of delivery and feeding[7,8]—and cannot be safely accelerated by means of diet or exercise.

Many practitioners of traditional "Closing the Bones" ceremonies—including Al Shedd, La Cerrada Postparto, and variations thereon—claim that such rituals contribute to the process of uterine healing. Massaging and/or wrapping the abdomen is said to support overstretched ligaments and muscles, readjust the uterus into its correct position within the pelvis, minimize bleeding, and even reduce infection. While there may not be a substantial bedrock of clinical evidence to support these claims, the fact that similar—in some cases, nearly identical—rituals evolved independently in such geographically and culturally diverse areas suggests that generations of women must have found some intrinsic value in these practices. It may be fair to examine or even critique this kind of lived experience, but hardly fair to ignore it.

Perhaps, beyond any tangible uterine benefits, these rituals remain popular because they offer the new mother something that has always been rare and precious: rest. Fatima Abdullah's voice is reverential as she tells me how women emerge from this peaceful state: "I give them time wrapped up," she says, noting that this gentle pause can last up to an hour, "and then I unwrap them. They usually come out of it very slowly, and they're like, 'Whoa. That felt great.' And it's like somebody just woke up from a very deep, relaxing sleep. The biggest reaction I get is that most people are able to turn off everything, not worry about anything, and just let their body have a moment of deep rest."

These traditional rituals may or may not have quantifiable effects on uterine healing, but the effect of "deep rest" on the postnatal womb and the person who owns it should not be underestimated. Rest from a crying baby, or from a fire that needs to be stoked, or from an email that needs to be answered, or from a pervasive pressure to do more, look better, heal faster, take up less space—anyone who has journeyed into matrescence knows that this rest has a value beyond measure.

WHAT WISDOM, THEN, HAVE WE ABSORBED FROM THESE practices that have evolved over millennia to serve and heal the new mother and her womb? Predictably, the industrialized Western world has distilled postpartum rituals into a form that suits its values: we have abandoned the mysterious, unquantifiable, unpalatably "foreign" elements of healing and have cherry-picked the one aspect—abdominal compression—that facilitates a return to a socially desirable postnatal body: a body that is slim, sexually available, and takes up as little space as possible. Mothering magazines, blogs, websites, and online retailers tout the miraculously transformative qualities of a new breed of compression garments designed to be worn for days or even weeks after birth. Variously known as belly bands, binders, wraps, waist trainers, and girdles, these items—often elasticated, sometimes adjusted by complicated systems of hooks, loops, and Velcro tabs, and sometimes even boned like the constrictive corsets of yore—are touted as essential tools in the new mother's armory.

We've already reflected on a moment of deep postnatal rest, a respite from the demands and expectations of a hectic world. Let's reflect now on a different kind of moment, one of vulnerability and insecurity: you are still a new mother, your body perhaps only hours or days removed from the exertions of birth, and it is the ungodly hour of night when no one but you and your baby seems to be awake. This child is tugging on your nipple, or perhaps gulping at a bottle, and as you ponder whether this is the fourth feed in an hour or just a continuation of one long, interminable guzzle, your uterus cramps and tightens. This is a normal part of postnatal healing—putting the baby to your breast initiates mini-contractions of the womb, helping to involute, a kind of biofeedback you remember your midwife mentioning—but the ache in your belly reminds you of the slack, burst-balloon skin that hangs in an untidy lip over your underwear. You pick up the mobile phone at your side, thinking that a one-handed scroll through social media might take your mind off your troubles, and a targeted ad flashes onto your screen: a pert, ponytailed woman in an elasticated wrap, standing sideways to show off the flatness of her belly. The website tells you that the band she's wearing "realigns your abdominal muscles" and "expedites fluids through the body," but another phrase interests you more: "reducing the appearance of 'mummy tummy.'"[9]

Something clicks as the baby continues its frantic suckling. It is as if someone has put a name to an itch in your brain, and offered to scratch it. You scroll on, you search, you click through a hallway of virtual windows. Another brand of belly band starts its spiel with the usual promise of clinical benefit, claiming that "such support bands are often recommended by physiotherapists to support the uterus, abdomen, pelvis and back," but a few lines down, you read a now-familiar refrain: "It's easy to see the belly bulge disappear and help gain your pre-pregnancy shape back by wearing a Belly Band."[10]

This promise is made by different brands in different ways all across the internet. Sure, each company bears its own disclaimer that these promises may not be fulfilled, but the fine print fades into oblivion as you read the testimonials of the seemingly endless happy customers.

"Omggggg. Girl. Yes. 5 stars!!" says one review. "I checked into the hospital on due date at 175 lbs. Gave birth and slapped this on the next day at the hospital. Came home the 3rd day looking like

Beyonce. Weighed 140 by the end of the week."[11] You scroll, you click, you find other women—women like you, with unruly wombs and sagging bodies—who write with the same breathless enthusiasm.

"I could already see my fat bulge disappearing after 2 weeks using my belly band," says one.

"My belly shrank to almost normal size so fast," says another. "Fingers crossed for a bounce back with this awesome waist wrap!" says a third.[12]

There are endless mirror selfies of women posing side-on, their bodies nipped into vast bands the color of sticking plasters—and in your hour of darkness, this goal is tantalizingly close: the snapback is within reach. Even the jelliest of bellies, the mummiest of tummies, can be tamed and constrained, almost as if you were never pregnant, almost as if your womb didn't exist at all.

In the interest of fairness, let's skip from the late-night feed to the lab bench. Could there be tangible benefits to the use of these garments? Can the claims of valuable support for pelvic muscles and organs be substantiated? In short, the evidence behind such claims is equivocal, and the evidence base itself is relatively minimal. Two small studies appear to suggest that the use of elasticated abdominal bands may decrease pain and enhance mobility in women who have birthed by Cesarean section.[13,14] However, the largest review of such studies to date found a lack of adequate evidence to demonstrate an overall effect.[15]

Belly bands' much-hyped promises to firm and restore the dreaded "mummy tummy" may be unfounded, too: Gráinne Donnelly, an advanced physiotherapist specializing in women's pelvic health, warns that commercially available garments are ill-suited to treat or correct *diastasis recti*, a postnatal separation of the abdominal muscles that can cause functional challenges and contribute to the appearance of a slack or bulging midriff.

She says, "I manage lots of postpartum women in relation to pelvic floor and abdominal wall recovery. Many women seek abdominal braces and compression garments for faster recovery and to help re-establish their prenatal figure. The reality of it is that the research supporting these garments and braces is lacking. In terms of abdominal wall recovery, the research does not support [the notion] that braces reduce or speed up recovery from diastasis. Research based on expert consensus released in 2019 advised that abdominal binders and braces only be used as an adjunct to rehabilitation in significant, pendular abdominal diastasis."[16] Women's health and fitness coach Beth Davies agrees, explaining that compression garments are "part of a toolbox, including restoring strength and coordination to the core/pelvic floor system, breathing patterns, management of intra-abdominal pressure, good nutrition and self-care."[17]

In short, there is no quick fix. "Mummy tummy" will not yield to bands or braces alone; a more comprehensive, expertly guided program is needed to reverse the stretches and strains of a full-term pregnancy.

Perhaps most crucially, it is reasonable to believe that the use or misuse of these binders in the postnatal period could lead to demonstrable harm. Many women find that urination and defecation can be slow to return to normality after birth; pelvic organs have been displaced and sometimes traumatized by surgery, forceps, catheterization, or prolonged pushing. As a midwife, I shudder to think what effect prolonged abdominal compression might have on these physiological processes: how can you break wind, or empty a distended bladder, or give your uterus the space it needs to contract and involute in its own time, when you are effectively squeezed into a human sausage casing for days on end?

One blogger tells her own cautionary tale of prolapse after some overenthusiastic belly binding: starting by layering several "hook and loop stomach wraps" over her abdomen at just five days postpartum, Jennifer Thomé says, "The compliments started rolling in. My stomach was flatter than it had been in years. Success! Or so I thought." This "success" was short-lived: increased lochia and a constant feeling of pressure led Jennifer to consult her obstetrician and midwife, who confirmed that she had a pelvic organ prolapse. "The bind squeezed all of my organs in," Jennifer writes, "which pushed my bladder down."[18]

In addition to the potential for physiological harm to the uterus and adjacent organs, one might easily ask what psychological harm these modern binding practices might cause. There may be many satisfied customers who look and feel like Beyoncé after a few days of binding, but after the initial euphoria of buttoning up your jeans just three days postpartum, what kind of lingering effect remains? A "mummy tummy" is evidence of an organ that has worked incredibly hard, and produced a new life. It may not be celebrated as a badge of honor, but if women are told that it is actually a thing of horror—a messy, ugly expanse to be hidden and compressed at all costs—what does this say about the way our society values or maligns the birthing body?

Against this grim backdrop of stigma and shame, Gráinne Donnelly strikes a note of cautious optimism; she thinks a time may come when there will be enough evidence to support more sophisticated, medical-grade garments in the treatment of specific postnatal health issues such as *diastasis recti* or incontinence.

"We will hopefully have more research supporting this in the coming years," she tells me. In the meantime, though, the vast array of products and the equally mind-boggling promises made by their manufacturers may be causing untold damage to a vulnerable population; enough damage, in fact, that Gráinne thinks the market for such products "should be regulated."

How, then, can the new mother and her womb be safely healed

and restored in a world that prizes form over function, socially constructed beauty over reproductive health? Fatima Abdullah says that there is growing interest in holistic postpartum traditions such as Closing the Bones and its many variations.

"I feel like there's actually a pretty nice resurgence in understanding that traditional forms of support have a lot of wisdom in them," she tells me. "People are kind of coming back to it." Perhaps, though, not every birthing person wants to be wrapped and released. Some might feel uncomfortable with appropriating or participating in traditions adapted from cultures not their own; some might simply lack the money or the time to be nurtured and anointed while there is a baby to be fed and work to be done. But surely every postnatal womb—after ballooning forth, and contracting down, and creating and expelling a whole new human (or two, or three, or more)—surely every uterus, having worked so hard, deserves a moment of deep rest, and permission to take up as much space as it needs.

Health

IN SICKNESS AND WELLNESS

I am always begging my body not to be so broken
but my body just laughs because it knows who started
this war.

—Fortesa Latifi, "Chronic Illness"

I am suspended six feet in midair, naked from the waist down with my legs spread and stirruped. The chair I am in has, according to the nurse who is now frantically punching buttons on a control panel, been "playing up" all week, and now it is on its absolute worst, most mischievous behavior. From my vantage point up here near the ceiling tiles, I can just about see the dark roots coming through at the top of the nurse's peroxide-blond head. I can see a cart with its drawers jammed full of cellophane packs: sterile speculums, swabs, needles, and syringes, all of the tools that may be needed to probe the secret pockets of my body. I can see the doctor, too, as she squirts a packet of lube onto the ultrasound wand that will soon be inside me; she glances up to my legs dangling above her, then at the increasingly apologetic nurse, her face slack with infinite boredom as her gloved hand rubs the lube up and down the wand. I watch it all from my perch up here in the air. I try to remember where I've seen the doctor's expression before, and then it comes to me—it's the weary resignation of the sex workers in Amsterdam's red-light district, pacing their windows in bikinis and pleather hot pants. This is another day in the slip and suck of women's bodies, and we're all just living in it, trying to do our best.

I crack a joke about being left high and dry; I am gracious; I try to defuse the situation, although I am the one with the least reason to apologize. In the other part of my life—the part where I am the one standing between other women's legs, cheering them on, willing their bodies to perform in the best, healthiest, most joyous way possible—I tell women not to keep saying sorry. You are wonderful, I say. You are beautiful and strong. Don't apologize for your body. But today, my flawed body has brought us together—the nurse, the doctor, and I— my body that labors every month to deliver those age-old twins, pain and bleeding—and I am saying sorry for my body, sorry for the chair, sorry you have to do this and see me and touch me and give a name and a cure to my problems, sorry, sorry.

There is no cure, though, the doctor tells me twenty minutes later in her office, once I have been rescued from my perch and positioned and probed in another room with a better, more obedient chair. She taps her pen against the computer screen on her desk.

"This is an ultrasound image of your womb," she says, tapping an area that looks like it's been blasted by a fine spray of grayscale frost, "and this part of your uterus is calcified."

Confused by the term, and still reeling slightly from my fairground-style experience in the first chair, I try to imagine what this diagnosis could mean. I have worked my fingers through the tight, chalky knots of calcified placentas, gritty giveaways of a chain-smoking mother, but I do not smoke, nor am I pregnant. I want to formulate an intelligent question, but all my clinical knowledge abandons me. Today, I am the patient. I am the woman who has been told to "take your bottom half off," and although I am now dressed, I feel disoriented and incomplete, as if I have put myself back together the wrong way. My head feels as remote from my body as a loosely tethered balloon. Clever words and jargon fail me; instead, I ask the question posed by patients everywhere.

"Is that bad?"

"No," the doctor says, avoiding my gaze. "It's just a change we would normally associate with the uterus of a much older woman. An aging womb."

So. It's fine. I don't have a terminal disease, my uterus is simply racing me to the grave. That strong, pulsing muscle—the organ that grew my daughters, nurtured them, and expelled them into the world—is, at the age of forty-two, turning to chalk, to the rigid, brittle stuff of bone and shell. This, perhaps, is why my periods have become so long and painful, with blood loss that makes my head spin: my uterus isn't supple enough to do its monthly business with ease. It struggles as it clenches—the healthy tissue strangled by its calcified shadow—it grumbles, it writhes, every cell straining to flex and relax. I am offered hormones and surgery, both of which I decline for reasons which I know to be valid, and yet still I apologize six more times before I gather my belongings and leave the doctor to type her notes in the scant minutes until my chair is occupied by another woman with another dysfunctional body. Later, as I walk through the hospital car park, I imagine my calcified womb rolling in my pelvis like a pebble, its richly veined fibers shot through with threads of stone.

While I now have an answer—if not a solution—to the monthly torment inflicted upon me by the organ that once served me so well, there are many reasons why a person with a uterus might experience pain—in and around menstruation, sex, movement, or at any other time—or troublesome bleeding. When I asked—when *we* ask—"Is it bad?" what we often mean is, "Is it cancer?"—our minds leaping nimbly to the worst-case scenario. Sometimes the answer is yes: womb cancer (an umbrella term for malignancies of the lining or body of the uterus) is the fourth most common cancer in women in the US and the UK, with over 66,000 and 9,000 diagnoses a year in these countries, respectively.[1,2] The prognosis for these patients—especially those whose disease was identified early—is often one of guarded optimism. Cancer Research UK reports that around three in four women diagnosed with uterine cancer in England survive their disease for five years or more, with survival rates rising to nearly nine in ten among the fifteen to thirty-nine age group.[3] Of course, no cancer diagnosis is ever welcome, but emerging research in this field—like that of Frances Byrne and her fellow uterine explorers—continues to enrich our understanding and improve outcomes.

Not all cancers are created equal, though, and not all wombs are afforded equal opportunities for diagnosis and treatment. Globally, cancer of the cervix—the neck of the womb—kills over three hundred thousand women per year or, to put it even more starkly, one woman almost every two minutes.[4] In fact, the disease kills more women worldwide than the usual uterine suspects, pregnancy and childbirth.[5] That statistic may come as a particular surprise to those who are fortunate enough to live in a higher-income country, with access to regular gynecological healthcare. After all, the "Pap smear"—the standard screening test for cervical cancer, named after its inventor, Georgios Papanikolaou—has become a rite of passage, albeit often an uncomfortable one, for anyone with a cervix in the wealthier, developed world. In the US and the UK, for example, cervical screening begins at age twenty-five and is repeated every three to five years thereafter.[6,7] The nervous jokes with the practice nurse, the cold slip of a speculum, and the odd sensation of a deep internal swab are all too familiar to many of us, but these moments of discomfort may result in the detection of a disease that is often asymptomatic in its earliest stages.

We actually have to thank Andromache "Mary" Mavrogeni

Papanikolaou, Georgios's wife, for this intervention; when her husband, a budding pathologist, discovered a new method of scraping cells from the cervix and studying them for precancerous changes, not only did Mary become his lab technician, but she offered her own cervix for repeated sampling.[8] "There was no other option for me but to follow him inside his lab," Mary later recalled, "making his way of life mine."[9] Mrs. Papanikolaou's generosity of body and soul may have been overshadowed by her husband's professional achievements, but her cervix has undoubtedly played a role in saving many thousands of lives since Georgios first published his findings in 1941;[10] regular screening with Pap smears has been proven to reduce deaths from cervical cancer by as much as 80 percent.[11]

In recent years, research linking the presence of the human papilloma virus (HPV) to cervical cancer has taken screening programs in a new direction. Previously identified as the virus responsible for genital warts, HPV is now known to cause the vast majority—a whopping 99.7 percent—of cervical cancers.[12] Given that roughly 80 percent of adults may be infected by HPV at some point[13]—often asymptomatically—vaccination against HPV now forms a key element of the World Health Organization's strategy to eliminate cervical cancer.[14] Screening still plays a part, too, albeit with a new, less invasive kind of smear test that detects HPV rather than specifically cancerous or precancerous cells. Crucially, self-sampling—in which people take their own "smears" or samples at home and then forward them to a lab for testing—has shown huge promise in recent studies. Information provided by the YouScreen study, a 2021 trial offering self-sampling to 31,000 women in London, states that "99 out of 100 people are able to do self-sampling properly."[15] The opportunity to take a simple swab in the comfort of one's own home may be transformative for people who might otherwise have avoided professional screening; for example, in a 2018 survey by Jo's Cervical

Cancer Trust, 72 percent of women who had survived sexual violence report delaying or declining screening.[16]

One survivor explains, "With smear tests and pelvic exams, your mind understands why these are necessary but often your body can not differentiate between the exam and the assault it experienced, so the smear test can feel both painful and like a violation to the body. I can say that if my vagina could speak when it comes to smear tests, it would be screaming, I do not want to go, LEAVE ME ALONE!!!"[17]

For some people (and vaginas) who have experienced assault, self-sampling may fulfill this desire for bodily solitude and autonomy while still providing an opportunity to participate in a potentially lifesaving intervention. Survivor or not, anyone who finds intimate healthcare encounters challenging or stigmatizing might well prefer this kind of screening; for example, among 137 trans and nonbinary participants in a 2021 study, 53 percent said they would like to be offered self-swabbing.[18]

The next challenge for the global health community is ensuring that screening and treatment for cervical cancer are available to people in less privileged nations. Cervical Cancer Action estimates, "In low- and middle-income countries . . . fewer than 20% of women have been screened for cervical cancer, compared with 60% in high-income countries,"[19] and the Global Surgery Foundation points out that even if access to screening is improved, "it is unethical to provide screening without effective treatment options."[20] In a more equitable future, every cervix—every neck of every womb—even those that have been abused, assaulted, or marginalized, or those that exist in developing or lower-income countries—will have an opportunity for optimal health.

—

ALTHOUGH MANY OF US IMMEDIATELY THINK OF CANCER when confronted by pain and bleeding, the truth is that these symptoms are far more likely to be caused by fibroids, endometriosis, or adenomyosis, conditions that may be experienced by millions of people with wombs every year, but which are seldom included in school-based sex and health education curricula. Growing up in America in the eighties and nineties, I sat through some version of these classes every year from sixth grade onward, dutifully learning the parts of a penis and the correct spelling of chlamydia and other "essential" facts, but the names of common gynecological conditions never crossed my teachers' lips. I was shown old-fashioned sanitary pads, with their medieval configurations of hooks and belts; I endured the ritual humiliation of rolling a condom onto a banana; I watched grisly films of drunk-driving teenagers lying bloodied and moribund at the roadside, but these "useful" lessons left gaping voids in my knowledge of myself. I was never told that my own body was likely to betray me, that the monthly cycle I plotted with my new tricolor ballpoint pen could bring its own trauma and hemorrhage, or that there were names for the things one's uterus could do—and, statistically speaking, probably would do—apart from having periods and babies.

Even as a student midwife, hungry for knowledge of all the inner workings of a woman's body, I only learned these things incidentally—endometriosis lesions observed during my one-day placement in gynae surgery, or fleeting descriptions of fibroids in a woman's maternity notes—and often had to deepen my understanding of these conditions by combing websites and journal articles in my own time. Occasionally, conversations with my midwifery colleagues—notorious for their candor—would reveal that almost all of us had suffered with debilitating pain and the aptly named "flooding" for years. Some of us had been lucky enough to have a diagnosis; others had been fitted with intrauterine devices and sent back out into the world without further discussion. Some—in fact, most—just put up and shut up. And we are supposed to be the educated ones, the ones with such deep and intimate knowledge of women's bodies that our power could once have seen us burned as witches. Even we—with our textbooks and anatomy charts and doctors on speed-dial—are often wandering in the metaphorical darkness with our wayward wombs.

As gynecological research gathers pace, though, and as it becomes more socially acceptable to discuss what was heretofore dismissed under the broad heading of "women's problems," shafts of light are appearing in that darkness. No longer happy to suffer in silence, people are talking about their wombs, and a diverse chorus of voices has joined the dialogue, from online forums to mainstream media; from the corridors of power to daytime TV.

–

"EVERYWHERE I GO IN THE WORLD," SAYS CYNTHIA Bailey, model, actress, and star of reality show *The Real Housewives of Atlanta*, "women are coming up to me, thanking me for having this conversation because it made them feel comfortable having the conversation."[21] Bailey isn't talking about the time she walked the runway in Milan, or her appearance in the music video for Heavy D's bombastically macho anthem "Nuttin' But Love." She's referring to the 2013 *Real Housewives* episode in which she opened up about her struggle with a condition that made her anemic, exhausted, and constantly fearful of an embarrassing public "accident." Bailey's candor struck a chord with her audience, and no wonder: uterine fibroids— the secret condition at the heart of this conversation—are estimated to affect as many as 70 to 80 percent of women in their lifetime.

Fibroids (or "leiomyomata" to give them their proper clinical name) are benign growths of muscle and fibrous tissue that can develop in and around the body of the uterus. They can be as small as your fingertip or as large as a melon, and they may occur singly or in the dozens. While some fibroids are completely asymptomatic, their presence only discovered incidentally during pelvic investigations for other conditions, many can announce themselves with ongoing, troublesome symptoms such as abdominal and back pain, heavy periods and bleeding throughout the menstrual cycle, painful sex, and increased frequency of urination. A 2015 study by researchers at Northwestern University found that the psychological effects of fibroids can be debilitating, too, with most of the sixty participants reporting fear, anxiety, anger, and depression related to the condition, and many describing a negative self-image that had interfered with intimate relationships.[22] These findings appear to have been confirmed more recently on a much larger scale: in a 2021 study of over nine hundred thousand women, those participants with fibroids were more likely to receive diagnoses of anxiety or depression. Unsurprisingly, this correlation was highest among women experiencing pain-related symptoms.[23]

Some of the data emerging from these studies are particularly poignant: half of the women in the 2015 cohort reported feeling "helpless," and participants with fibroids in the 2021 study were found to have a higher rate of "self-directed" violence than those without. These statistics tell a tragic tale of desperation, and of women struggling to cope with an uncertain future in whatever way they can. Speaking to *Essence* magazine in 2020, fibroid awareness activist Tanika Gray Valbrun explained the relentless stress of attempting to manage her frighteningly unpredictable condition: "I developed an intense anxiety about messing up car seats and mattresses or standing up in a business meeting and feeling the telltale gush . . . I never wore white, which was an emotional reminder of how fibroids controlled my quality of life."[24] This theme of fear and frustration—of not knowing when pain or bleeding might strike, of coping in silence and shame, of not seeing a clear path forward—is echoed time and again in studies of women with uterine problems of any origin, and a sad reflection of the persistent gaps in our knowledge of life-altering gynecological conditions.

While there is no way to prevent the occurrence of fibroids, there are some clear risk factors that contribute to an increased incidence: older age, obesity, a family history of the condition, red meat intake, vitamin D deficiency, and even a history of child sexual abuse have all been linked to the condition.[25] The greatest risk factor, though, is one which is impossible to alter: race. A growing body of evidence suggests that Black women are two to three times more likely to develop fibroids during their lifetime, with as many as 90 percent of Black women experiencing one or more fibroids by the age of fifty. Black women also tend to develop fibroids earlier than their white counterparts, and these growths tend to be larger, more numerous, and more severely symptomatic.[26,27] While these symptoms can sometimes be managed with treatments ranging from oral anti-inflammatories and anticoagulants to hormonal medications to more invasive procedures such as myomectomy (surgical fibroid removal), Black women are twice as likely to undergo the last-resort "cure" of hysterectomy: removal of the uterus itself.[28] Whether this last fact is a product of genuine clinical need or a wider societal devaluing of Black people's reproductive lives is a question that continues to generate fierce speculation and debate.

Stephanie Tubbs Jones, a state representative from Ohio (and the first Black woman to be elected to that role), argued that "women deserve better" than to suffer needlessly from this potentially debilitating but often undiagnosed condition. "Women," she wrote in 2007, the year before she died, "we can no longer be silent about what ails us. If we don't speak out, our silence could be our downfall."[29] Recognizing the havoc wrought by fibroids on the lives of family members and friends, in 1999 Tubbs introduced a congressional bill to increase funding into fibroid research. While the bill floundered and Tubbs died without seeing this campaign come to fruition, her work has been resurrected in a similar bill proposed by New York representative Yvette Clarke and supported by Vice President Kamala Harris. The 2020 Stephanie Tubbs Jones Uterine Fibroid Research and Education Act—possibly the first American legislation to feature the womb in a starring role without any relationship whatsoever to abortion—would provide an annual stipend of $30 million for fibroid research, enhanced data collection around the condition, and the creation of a public education program. At the time of writing, this bill is making its slow but steady way through congressional due process; by the time of publication, it may be enshrined in law, initiating enough public awareness of uterine fibroids that women may no longer have to learn about the condition from a chance Tuesday-afternoon channel-surf through *The Real Housewives of Atlanta*.

—

PERHAPS THE MOST COMMON DISORDER RELATING TO the uterus, and the one that can have the furthest-reaching, most comprehensive and devastating effects on a person's health, is the one that is least understood, hardest to diagnose, and most challenging to treat: endometriosis. In this condition, tissue similar to that of the endometrium, the womb's lining, adheres to and grows around structures throughout the body, from the neighboring bladder and the bowel to the more distant lungs, the liver, and even the eye. As estrogen and progesterone wax and wane throughout the month, thickening and shedding the tissue within the womb, so, too, do these fragments of endometrium swell and bleed, in turn causing pain and distress that can bleed into every area of the sufferer's life.

"[My uterus] appears normal," writes the actress and author Lena Dunham in the March 2018 edition of *Vogue*, "cheerful in blonde pigtails like little Rhoda the evil child from the classic film, but it's angry, exhausted, a home for no one."[30] With raw anguish and frustration, Dunham describes the myriad ways in which her womb has obliterated her physical and—perhaps even more painfully—mental health. It's no wonder she personifies her uterus as a kind of schizophrenic devil child; once an organ that held the promise of a longed-for baby, now the cause of dashed hopes and broken spirit. Dunham explains her decision to have a hysterectomy; for her, after years of misdiagnosis, suffering, and failed treatments, a body without a womb is the only cure for a womb that's invaded every part of her body. Although she rose to fame for writing about the gritty, glamorous lives of her fellow millennials, in latter years, Dunham has become the reluctant poster girl for a devastating disease—the one so doggedly researched by Dr. Christine Metz in her ROSE trial, and still so little understood by the medical community at large.

Sometimes described as a disorder of early embryological formation, widely characterized as a gynecological disease, now variously regarded as an inflammatory, metabolic, or even neurological condition, and still sometimes dismissed as a figment of the sufferer's imagination, endometriosis—or "endo" as millennial hashtags would have it—is estimated to affect as many as 176 million people worldwide.[31] Some figures point to as few as 2 percent of the world's population, while others suggest that endometriosis may be found in as many as five times that number. These varying estimates are indicative of endo's elusive nature; with diagnosis currently only possible by means of costly, invasive procedures, many women endure the condition for an average of seven to ten years before receiving a definitive diagnosis. The true number of people with endometriosis may indeed be much higher, as those without access to healthcare, or less likely to engage with the system for reasons of financial or social disparity, may yet be suffering in silence. Even for those women fortunate enough to engage with comprehensive, effective healthcare, there is no absolute treatment for endometriosis, and even hysterectomy—the go-to final fix for so many gynecological conundrums—is hardly a cure for the lesions that can also be found in far-flung corners of the body.

Doctors have been hot on endo's heels for centuries, with descriptions of similar lesions found in medical texts as far back as 1690. Perhaps the most recognizable early depiction of endometriosis comes from William Wood Russell, a doctor writing in the Johns Hopkins Hospital Bulletin of 1899. Russell describes how, on examining a patient's right ovary, "we were astonished to find areas which were an exact prototype of the uterine glands and interglandular connective tissue."[32] The challenge to define this mysterious syndrome was taken up with enthusiasm by John Albertson Sampson, a New York gynecologist who is often referred to with patriarchal flair as "the father of embryology." Sampson first coined the term "endometriosis" in 1925, and two years later he published his "backflow" theory that still pervades the study of the field to this day.[33] Perplexed by the presence of what looked like menstrual tissue outside the womb itself, Sampson theorized that this tissue could only have migrated out of the uterine tubes and into the pelvic cavity in a kind of reverse, or retrograde, menstruation. Sampson's contemporary, Baltimore-based gynecologist-pathologist Emil Novak, was highly skeptical of this theory, casting doubt on the "current" that would be required for this kind of backflow and noting that he himself had never witnessed any blood within the pelvis of any immediately postmenstrual women during abdominal surgery. Throwing the kind of passive-aggressive shade often found in polite scientific debate, Novak said of Sampson's theory: "This would seem incredible."[34]

Nevertheless, the paradigm of retrograde menstruation persisted until the 1980s, when David Redwine, a doctor working in solo private practice in Oregon, turned this theory on its head. Even the name Redwine is evocative—the countless tiny deposits of endo scattered throughout the body like droplets of wine spilled across the table from a toppled glass—and Redwine theorized that this scattering of tissue occurred in the earliest moments of human life. According to his novel theory of "Mulleriosis," endometriosis lesions are the result of a glitch in the development of the Mullerian ducts, those primitive structures that regress in males and evolve into a urogenital tract in females. Redwine hypothesized that some Mullerian cells might differentiate and then migrate during embryonic development in such a way as to lay down womb-like tissue in the wrong parts of the body, setting the stage for the pain and distress of endometriosis in later life.[35]

This theory appears to have been borne out by research: in one study of stillborn female fetuses and another of female infants who died in the neonatal period, endometriosis tissue was identified in 11 percent of subjects.[36] In 2015, a remarkable case study from Pennsylvania describes a thirty-five-week fetus with a large mass in her abdomen that, during neonatal surgery, was found to be a hemorrhagic endometrioma; in other words, a massive, blood-filled collection of endometriosis.[37] Given the fact that female fetuses have never experienced menstruation, Sampson's backwash theory seems to be less and less plausible, and Lena Dunham's depiction of her uterus as "little Rhoda the evil child" may, in fact, have its origins in the first weeks of embryonic girlhood. To further complicate matters, ongoing research from Edinburgh University's MRC Centre for Reproductive Health indicates that the broadly defined disease of endometriosis may actually have three distinct subtypes—cystic ovarian, superficial peritoneal, and deep—that could be more effectively managed using type-specific treatments.[38]

In time, a deeper understanding of these subtypes may lead to more personalized and effective therapies. Until then, even the "model" endo patient—one who is educated and informed about her condition, and keen to advocate for herself—continues to navigate a medical system that can be confusing at best, and dangerously undermining at worst.

The road to diagnosis and even moderately effective treatment can be fraught with obstacles, and the emotional toll of the disease (and the ensuing battle for validation) can be scary, exhausting, and downright overwhelming.

Michelle Hopewell, an actor and endo awareness activist from Edinburgh, watched her mother endure years of pain and treatments of varying success before she herself developed symptoms of the disease.

"I grew up around endometriosis," she tells me, "because my mom had stage four [the most severe and widespread form]. Before we actually had a name for it, my parents went on this very long journey to get my mom a diagnosis because she always had heavy and painful periods. By the time she had me, her health just declined. This was twenty-plus years ago, and at that point, there was even less understanding than there is now, especially here in the UK. My dad started reading up on it, and he and my mom were just trying to figure out what it might be. They'd gone to Harley Street in London, you know, they were paying out of their own pocket, and whether it was in the NHS or privately, they were being told things like my mom had psychosis, and maybe it was phantom pain, and even if she did have pain, Black women have higher pain thresholds and she should just try to deal with it. There was all this really intense medical racism coupled with a severe lack of knowledge and understanding in the UK at that time. So they ended up going to America and they saw a gynecologist in Georgia who was a specialist in endometriosis. He finally diagnosed her and that's how we came to know within my family what endometriosis was."

Although Michelle can often be found smiling through her endo flares on social media, sending messages of love and positivity to the followers she calls her "Great Lights of the Universe," her weariness is evident as she describes this ominous presence in her upbringing, like an unwanted, unruly sibling. She grew up alongside her parents' battle with endometriosis, and, as it turned out, endometriosis was growing inside her as well.

"By the time I was having periods at age ten or eleven," Michelle recalls, "my parents were old hands with endometriosis." Cycle after cycle, Michelle struggled with the same pain she'd watched her mother endure for years, but the trauma of witnessing her parents' conflicts with the medical establishment had already taken its toll. Suffering and denial were Michelle's main coping strategies until the severity of her symptoms drove her to the doctor. "I didn't like going to the GP and I didn't like being unwell or sick. It took me until I was around twenty to admit to myself that that's what it could be, and it still took me until I was maybe twenty-three or twenty-four to start going to GPs to investigate it. I'd do everything in my power to avoid it, so what I did was just endure the pain. It was like a dissociation thing for me: I'm someone that hustles and I carry on, and I just pushed through and I persevered. I wouldn't allow myself the space [to suffer]. Even if I was at my most unwell, I'd still turn up and do it," Michelle says, recalling a tour of the musical *Matilda* during which "there would be days where I'd be onstage and genuinely feel like I was about to pass out. And so I would push to the end of the scene and then be off in the wings and on the floor."

When Michelle finally forced herself to embark on the same quest for a diagnosis that had dominated her mother's life for so many years, she was met with similar reactions of disbelief and dismissal.

"A lot of doctors were like, oh, it's your age, or you know, it's just how it's meant to be," she recalls. "And all the time, I'd flag my mother's history, and say, it's hereditary, we could catch it now. And I was constantly kind of gaslit, to be honest. And then I'd have to garner the courage to go back each time, and each time it gets harder and harder because you're fighting a battle—a battle I've seen fought before, that was very hard—and you just don't know if you have the strength to repeat patterns. Literally from a child to an adult, I'm often dismissed. And my two older sisters say they have the same experiences, and many other Black women and people echo the same stories. You're sometimes made to feel like you're an addict or something, like you might be coming with an ulterior motive. You're playing against stereotype; there's the fear of being seen as the Angry Black Woman: someone who's too much. And you end up having to walk a fine balance of trying to get the help you need, because you're often disrespected regardless of what you say or do. This person will see you and have a multitude of preconceived notions that you can't really do anything in that moment to fight against, you just have to hope that they want to do their job. That's something that's very scary for me, because there's an element of feeling like, if I'm really in trouble, is there anything that I can do or say that will get me help?"

Recent research indicates that Michelle's struggle to be heard, validated, diagnosed, and managed effectively may not be unusual, especially among other women of color. In 2019, a Canadian team's meta-analysis of twenty clinical studies found that Black women were less likely than their white counterparts to be diagnosed with endometriosis.[39] However, this discrepancy does not necessarily mean that Black women are less prone to the disease; it may, instead, be indicative of some practitioners' implicit bias. A subsequent commentary on the Canadian review notes that, for many years, endometriosis was erroneously believed to afflict mainly white, upwardly mobile, professional women; the authors suggest that these outmoded ideas "may still consciously or unconsciously influence clinical care."[40] In light of the difficulties experienced by Michelle and her mother, sisters, and friends, this cautious doubt appears overly generous; these women's stories suggest that racist attitudes—whether implicit or explicit—absolutely affect and impede Black endo patients' journeys through the healthcare system.

Sadly, the theme of being disbelieved or "gaslit"—having one's own lived experience questioned, and another version of the truth superimposed upon it—appears time and again in endo narratives from women of all walks of life. In *Ask Me About My Uterus*, her memoir of living with the often crippling and all-pervasive effects of endometriosis, Abby Norman writes, "Even outside of the doctor's office and in social settings, women face a constant barrage of doubt that undermines their faith in their own internal experiences. They begin to question their reality."[41]

It's little wonder, then, that—as with uterine fibroids—there is increasingly strong evidence of a relationship between endometriosis and poor mental health. In a 2019 survey of over 13,500 women, co-published by the BBC and Endometriosis UK, 50 percent of women with endometriosis reported suicidal thoughts, and most respondents said that the disease had had an adverse effect on their education, their career, and their relationships.[42]

Lauren Mahon, a health activist, describes a typical endo journey and the ensuing effects on her mental health: "I felt like a failure because I didn't get diagnosed for five years. I thought I couldn't handle period pain. I'd go to the walk-in clinics and go, 'Something's wrong with me,' and they'd feel about and do stuff, they'd do a scan. You can't see endo on a scan, so they'd be like, 'We can't see anything wrong,' and I just felt like a failure. I couldn't have sex with my boyfriend [because of the pain]. I nearly lost my job because of all my sick days, and I'm sat there as a 24-year-old woman thinking it's all my fault."[43] For Mahon, the suspicion, disbelief, and lack of sympathy from the important people in her life made her journey with endometriosis nearly as traumatic as her subsequent illness with breast cancer. When your employer, your partner, and even your medical caregivers discredit your symptoms and dismiss the pain caused by these tiny spots of rogue tissue, it's easy to wonder if you are, in fact, going mad.

One 2017 study of doctors' attitudes toward women with endometriosis appears to confirm a clear mistrust of these sufferers' voices, and an assumption that the disease is simply a physical manifestation of an already dysfunctional mind. "Do mad people get endometriosis or does endo make you mad?" one gynecologist in the study is quoted as saying. "It's probably," the doctor concludes, "a bit of both."[44]

WHILE IT SHOULD BE NOTED THAT MANY WOMEN DO have satisfying and therapeutic engagements with mainstream modern gynecology, the voices of Lauren Mahon, Michelle Hopewell, and others like them form a chorus too loud to be ignored. Working against such a deeply embedded skepticism around women's lived experiences, and a mistrust that's all too often amplified by entrenched racism and bias, it's no wonder that so many women find their mental health dictated, in large part, by their uterine health. This connection is not new; its roots are insidiously deep. From the earliest recorded history, men and medics (and, for millennia, the two were one and the same) have argued that women's moods and minds are controlled by their dysfunctional wombs. Since the dawn of civilization, almost as soon as man could put pen to paper (or quill to papyrus), he began to document his impression of the uterus as a wily, wandering, problematic organ. In much the same way as we might now personify a particularly nasty virus, the womb has long been seen as a mischievous marauder that can whip through the body on a whim, wreaking havoc on every organ in its path until finally, it infects the patient's brain. The learned onlooker strokes his beard and nods sagely while the owner of said uterus displays all the classic hallmarks of madness: delirium, seizures, hallucinations, or just an inconvenient difference of opinion. As recently as 2015, the so-called leader of the Western world made a very public comment associating women's whims with their wombs. Irked by what he perceived as an attack by journalist Megyn Kelly during a televised debate, then president Donald Trump later insinuated that Kelly's unfavorable behavior must have been caused by a kind of all-consuming menstruation: "She gets out there and she starts asking me all sorts of ridiculous questions," Trump railed, "and you could see there was blood coming out of her eyes . . . blood coming out of her wherever."[45] Women, wombs, and their messy, mad "wherevers": a stigma as old as time, and as prevalent as ever.

We may look back and laugh at the notion, expressed in the *Ebers Papyrus* in 1600 BCE, that a woman's wandering womb could be tempted back into place by massaging her vulva with sweet-smelling unguents—a kind of early uterine catnip. Another Egyptian papyrus dated to the third or fourth century contains an incantation to cure such uterine disarray:

> *I conjure you, O womb . . . that you return again to your seat,*
> *and that you do not turn [to one side] into the right part of the*
> *ribs, or into the left part of the ribs, and that you do not gnaw*
> *into the heart like a dog, but remain indeed in your own intended*
> *and proper place.*[46]

On reading such a prayer—which the supplicant is instructed to write on a "tin tablet" and "clothe" in seven colors—one might easily scoff at such a primitive model of health, and wonder at the naive simplicity of such a cure. And yet, for some reason, this notion of the most uniquely female organ as the origin of disease—especially psychiatric disease—persisted for thousands of years. Doctors, philosophers, poets, and presidents have long blamed the uterus for erratic female behavior. In fact, they're still doing it today.

It would be easy to dismiss this uterus-brain connection as simplistic, misogynistic, or both. How ridiculous to think that the uterus has a hotline to the brain, tugging on it like the wire from a scullery maid's bell in a stately manor. How infuriatingly crude to imagine that every woman's behavior can be reduced to a hormonal glitch, and that she navigates life like a lab rat staggering blindly through a maze, a slave to her basic biology. As we'll see, the truth is far less ridiculous or crude, but the relationship between the womb and the brain is too plausible—and potentially game-changing—to be dismissed.

First, it's important to understand the origins of our modern ideas about the womb's effect on mental health. If we imagine the collective consciousness as an onion, with the newest outer layer as a smooth, shiny skin of progressive equality where women are completely in control of their own minds and destinies, and the innermost, oldest nub as a kernel of cave-dwelling ignorance, then it's fair to say that the intervening layers of our historical onion are deeply imbued with the sharp, acrid tang of hysteria.

"Hysteria"—from the Greek word for uterus, *hystera*—has been used for millennia as a kind of catchall term for female behavior that doesn't fit the socially or medically decreed norm. Hysterical women have historically been seen as uncontrollable, unpredictable, and dangerous, often with sexual perversions, unseemly thoughts, and even magical powers that are at odds with mainstream ideals of femininity. Some of the symptoms attributed to hysteria over the years would today be seen simply as quirks of personality or, in some cases, ascribed to diseases as diverse as epilepsy, bipolar disorder, anorexia, anxiety, depression, chronic fatigue syndrome, and fibromyalgia. However, it took us a while to get to this "enlightened" position. For many hundreds of years, men (and it really has been mainly men who've had the privilege of establishing the medical zeitgeist) have studied the behavior of women in the name of science; prodded and paraded them, "treated" them with potions and punishments, and finally shrugged their shoulders and sighed, "Hysteria." And for many hundreds of years—right up until 1980, in fact—hysteria was a legitimate, respected, and widely recognized diagnosis.

Let's go back, though, from the gray, grim days of miners' strikes and spiral perms, to the sunnier climes of ancient Greece, where the idea of hysteria originated. Women had clearly defined roles—wife, daughter, slave, servant—and any female attitudes that challenged contemporary norms were often attributed to reproductive dysfunction. In Greek mythology, for example, soothsayer and healer Melampus is said to have lamented that the local Argonaut virgins "refused to honor the phallus and fled to the mountains." One can only imagine what kind of phallic temptations—or threats—caused these women to flee en masse to the hills, but this denial of masculine charms was deemed by Melampus to be a sure sign of "uterine melancholy." According to him, this disorder was caused by a lack of orgasm and, as such, could only be cured by copulation with virile young men. Oh, and hellebore, too—a handy herbal supplement in case penises didn't quite do the trick.

Plato, Aristotle, and Hippocrates also riffed on this theme of sex, wombs, and madness, with the latter coining the term "hysteria" and integrating it into the belief that all health was a function of four essential "humors." Blood, phlegm, and yellow and black bile were thought to control physical and psychological well-being, with variations depending on whether the body was wet, dry, warm, or cold. Women, Hippocrates wrote, were cold and wet; men, warm and dry. He believed that sex could bring the innately clammy woman back into balance; conversely, abstinence could lead to a "putrefaction of humors" in the unfortunate female. It stood to reason, then, that the best prevention (and cure) was for all women to be married, thus enjoying a healthy sex life while upholding social norms. And as for the odd singleton or spinster who continued to deny the charms of the phallus? Hysterical.

These beliefs were further articulated by Claudius Galen, a second-century medic who earned his early stripes as the official doctor of a gladiatorial school in Pergamon and went on to serve three Roman emperors. Galen argued that not only did women need men's sperm to heat their blood and open their internal passages, but they could also be afflicted by *furor uterinus*—literally, "womb fury"—if this insatiable need was not met.[47] Again, the cure was sex, with a few herbs thrown in for good measure. Quite convenient, one imagines, for the lusty fighters and kings who paid Galen's wages.

We can only guess what the good doctor's female patients felt about their diagnosis, as there are no firsthand accounts from the women afflicted with womb fury. While it's appealing to imagine that they were too busy marauding angrily about the streets of Rome to jot their thoughts down in a diary, the truth is that the voices of women—especially those deemed mentally unstable—are far outnumbered by those of men in the historical record.

Wayward women continued to be a hot topic for medical men over the following centuries, though, and treatments for this kind of problematic behavior are well documented. Some cures were undeniably foul, like seventeenth-century French physician Lazarus Riverius's "uterine elixir" of herbs and spices simmered with horse dung and wine.[48] Others were professionally dubious but potentially pleasurable, like the remedy suggested by Riverius's Italian contemporary Giovan Battista Codronchi. According to the good *dottore*, midwives could treat hysteria by manually bringing a woman to orgasm, thereby encouraging a healthy production of semen (thought, at the time, to be a female emission).[49] As the Reformation swept across Europe, the lines between hysteria and witchcraft became increasingly blurred and the "treatments" increasingly lethal, with hanging, stoning, and drowning taking the place of herbs and potions. Being a woman—

and owning a womb—had become decidedly dangerous in a world where the slightest deviance from the norm could be seen as not just mischievous or inconvenient but downright satanic.

Finally, in a Parisian asylum in the nineteenth century, a man with a fiery temper and a pet monkey called Rosalie decided to explode the myths of hysteria once and for all. Jean-Martin Charcot (a doctor, not a patient, in spite of his personal eccentricities) devoted his life to this most mysterious of female maladies, and his clinic at the Salpêtrière hospital became a world-renowned focus of the field. Charcot achieved professional acclaim and a certain degree of popular fame by staging theatrical demonstrations in which his ragtag group of "hysterics"—women who invariably had led lives of poverty, abuse, and vulnerability—could be provoked into bosom-heaving seizures, seemingly impossible contortions, erotic hallucinations, and dramatic swings of distress and ecstasy. These shows confirmed Charcot's reputation as an iconoclast, and his "modern" ideas about hysteria were equally unconventional. With a consuming interest in the emerging field of neurology, Charcot believed that hysterical behavior had its origins not in the uterus but in the central nervous system.

At last, you say. A man who believes that women aren't just slaves to their sex organs. Disappointingly, though, Charcot wasn't quite the forward-thinking neurologist one might have hoped for. Although he questioned the physical origin of hysteria, he also believed that hysterical episodes could be triggered and relieved by pressure on the ovaries, and he even designed a kind of gruesome turn-screw girdle for that very purpose. The uterus may have been off the hook, but Charcot just moved the blame for hysteria a few inches to the left and right. The female reproductive system still, in his model, had a lot to answer for. To twenty-first-century eyes, so did the doctor himself; his relationships to the women in his care might charitably be described as complicated, and more accurately described as exploitative and abusive.[50] Certainly, they were questionable enough to fail even the most fleeting scrutiny of a modern-day ethics board.

Charcot may not have completely discredited the idea of hysteria, but his work frayed some important threads in the already tenuous connection between the uterus and the brain. When, a few years later, Sigmund Freud suggested that the condition might be caused by adverse childhood experiences, and then theorized that even men could have the disease, the idea of hysteria's uterine origins began to fall from fashion. Twentieth-century advances in psychology, psychotherapy, and neurology saw this fall continue until finally, in 1980, the term was removed from the American Psychiatric Association's *Diagnostic and Statistical Manual of Mental Disorders* (*DSM*)— the bible for students of the mind and its ailments.

–

THANKFULLY, SCIENCE HAS NOW DISPROVED THE THEORY of the wandering womb, hysteria is no longer a recognized psychiatric disorder, and contemporary ideologies seldom favor such a reductive concept of female identity. However, as Abby Norman wrote in her endometriosis memoir, many women continue to find their identities poorly reflected and their needs inadequately met by conventional medicine.

"You're just banging your head against so many things," Michelle Hopewell says, "and you kind of have no choice but to look in any other direction that you can, so holistic health and wellness has definitely become something very big for me in terms of managing my pain."

For Michelle and many people like her, wellness—at once a nebulous and evolving concept and a multibillion-dollar industry—has rushed into the vacuum left by mainstream reproductive healthcare providers. In this brave new world of wellness, the uterus is identified as the source of myriad troubles and toxins, while also celebrated as the seat of a divine and intuitive female identity. Today's frustrated patient may find the doors of medicine slammed literally or metaphorically in her face, but she need only open her internet browser to find countless self-proclaimed "experts" only too happy to promise total healing of the whole woman, starting from her womb and working outward. The price tag for such remedies can vary widely, from a few dollars for herbs from a Bronx *botánica* to tens of thousands for a regular regime of personalized coaching and supplements from one of the self-appointed wellness elite.

Of course, given the historical preoccupation with women and their problematic wombs, quirky cures for uterine woes are not new. We have already encountered leeches, cupping, and foul-smelling herbs and unguents; see also the "hysteric juleps" (mocktails of castor, black-cherry water, pennyroyal, and henna) from the *Ladies Dispensatory* of 1739, and "Glanoid Multigland Liquor," an animal hormone extract sold—perhaps as a conveniently commercial by-product—by a London meatpacking firm in the late nineteenth and early twentieth centuries.[51] It seems as though everyone and his brother (or, in the case of Glanoid, even his butcher) has an opinion about the uterus and how to fix it, but one man, John Kellogg, is often credited with originating the wellness industry as we know it today.

Many accounts take a tone of affectionate bemusement when referring to the eccentric techniques espoused by Kellogg, inventor of the famous cornflakes and proprietor of Michigan's Battle Creek Sanitarium. Touted as a restful and restorative escape for forward-thinking Americans at the turn of the twentieth century, Kellogg's estate offered "cures" such as vibrating chairs, electric sunbeds, and industrial-strength enemas, but his endorsement of brutal treatments for women's health is less well known. Kellogg's 1892 *Ladies' Guide in Health and Disease: Girlhood, Maidenhood, Wifehood, and Motherhood* opens with a full-color illustration of "the human body" although, somewhat unhelpfully, the body in question is male. The rest of the book demonstrates a similar disregard and, indeed, an almost malicious distrust of the female body; Kellogg blames the uterus and its reckless owner for pretty much everything. Poor muscular development, masturbation, abortion, and birth control (including early withdrawal and the use of contraceptive "womb veils") are castigated as the root causes of ailments from prolapse to cancer. For a cure, Kellogg suggests "pledgets" (vaginal pessaries) soaked in vinegar and boric acid; enemas brewed from "pancreas and cream"; and the application of carbolic acid to the clitoris (or even removal of the clitoris entirely). Somehow, these brutal treatments don't feature in the 1994 film *The Road to Wellville*, a fictionalized account of a stay at Kellogg's sanitarium, and Kellogg is still often referenced as a quirky wellness pioneer, rather than the violently misogynistic quack his treatise would suggest.

The term "wellness" as we now know it was first introduced in the 1970s, when John Travis, a resident doctor in preventive medicine at Johns Hopkins University, developed a tool he called the Illness-Wellness Continuum.[52] On one end of this continuum is premature death, and on the other, high-level wellness, with the middle ground occupied by a spectrum of awareness, education, and growth. Travis argued that traditional Western medicine simply aimed to move a person toward the middle, a neutral point at which health is seen merely as the absence of disease—whereas true wellness meant a constant journey toward optimal physical, mental, and emotional fulfillment.

At the same time, second-wave feminists began to develop their own version of wellness, with the womb and its sexual and reproductive functions at the core. Perhaps the best written example of this radical new approach to health is *Hygieia: A Woman's Herbal*, published in 1979 by midwife and activist Jeannine Parvati, then a master's student in San Francisco. Parvati steers her readers away from medicalized knowledge, guiding them instead toward a universal and deeply embodied female truth: "For too long we have deified reason, to the expense of intuition."[53] Acknowledging the dominance of male gods and healers in the ancient Greek tradition from which early medicine emerged, Parvati suggests a new model based on the goddess Hygieia, daughter of Asclepius, the god of medicine. "The male healers, at least in the Greek tradition, had better press than the woman healers," Parvati writes. "And so it is to Hygieia that we draw our attention. She is the goddess within each of us who knows the grace of being healthy."[54] Parvati follows this statement of intent with a comprehensive guide to herbal cures for almost every aspect of reproductive health, from irregular periods to pregnancy, birth, and menopause. At the heart of every tea, tincture, and poultice is this radical new respect for the womb, an organ Parvati defines as "every person's first home" and "a woman's word for her source of power within her centre."[55]

Over forty years since Parvati urged women to locate their power source, the wellness industry—and specifically what I will call the "womb wellness" industry, a particularly profitable and problematic submarket—has mushroomed into one of the world's most lucrative commercial sectors. Consultancy firm McKinsey estimates the current value of the global wellness market at $1.5 trillion—that's right, *trillion*—with an annual anticipated growth of 5 to 10 percent. McKinsey describes the industry as incorporating a variety of products, from books and supplements to retreats, personal coaching, digital trackers, and noninvasive treatments.[56] Many of these products are marketed as solutions to nongynecological problems, but a substantial (and growing) number of them are indeed aimed at consumers engaged in the age-old battle to understand and control their uterus.

Purveyors of womb wellness tend to fall into one of two camps: on one side, the earthy, grassroots inheritors of Jeannine Parvati's legacy—often solo practitioners of complementary therapies or purveyors of small-batch herbs, pessaries, and potions—and on the other, manicured gurus with megawatt smiles and sophisticated websites selling big-ticket "merch" and expensive membership to online communities. The latter camp are perhaps more prominent in the public eye, with Gwyneth Paltrow and her online blog/marketplace, Goop, achieving equal parts fame, infamy, and fortune for their endorsement of eclectic and often costly treatments. Since Goop's inception in 2008, many would-be wellness gurus have surfaced in the website's wake. These women can be easily identified by their common characteristics: often, the "expert" in question is slim, white, glossy-haired, and conventionally attractive; entry-level purchases of supplements that promise to "balance hormones" and ease gynae complaints are often gateways to more costly "programs" that include semipersonalized coaching or membership in exclusive groups. Closer scrutiny of the expert's promises often reveals weight loss as an incidental (but whoops, oh-so-appealing) side effect of optimal uterine health. While Silicon Valley tech bros make the news for biohacking their way to productivity and virility, the womb-owners of the world—too often disappointed by their adventures in mainstream medicine—are hemorrhaging money into clinically questionable biohacks of a uterine kind, hoping to "cycle-sync" and "flow" their way to health and harmony.

It's easy to understand how the womb wellness industry has become so profitable in recent years. These programs are seductive, with their beautiful, charismatic leaders and their promises of transformative healing. To some people, like Alice, a student midwife from Manchester, the feeling of belonging that comes from being part of an online community, and the perceived physical benefits of following an alternative expert's advice, are worth the hefty price tag. Having read a menstrual health manual by one such expert, Alice signed up to the author's online "collective" and now has access to forums, an app, and individualized coaching.

"It's really hard to pick one thing I like the most, but the community is as valuable as the content shared. Whether it's listening to someone being coached on a call, or seeing a relatable post within the app, there's so much to gain from hearing someone's story, and sharing our own, and feeling the love back."

This "love" comes at a cost of £69 per month in the UK, an expense that didn't sit well with Alice at first. "I only intended to stay for a few months," she says, "because I felt bad spending so much money on myself every month, but I quickly saw results and the value of investing in myself."[57]

Nicola Goodall, an Edinburgh-based doula and traditional healer, could not be further from the immaculately curated image of the mainstream womb wellness practitioner. With her hijab and hoop earrings and an Instagram account as likely to feature old-skool rap as antiracism training days, Nicola has spent the better part of twenty years providing holistic treatments from womb massage to birth support, herbal tinctures to aromatherapy oils. She is well aware of the forces driving women away from industrialized gynecology; she witnesses the fallout of these futile engagements in her daily practice, and her services are in high demand. "Mainstream medicine as a whole has not been concerned with the well-being of women apart from their capacity as breeders," she says. "For a very, very long time, Western medicine has not given a shit about the womb apart from it being a container for the next male heir."

Considering the harmful legacy left by misogynist medics, from Soranus and Melampus to Charcot and Kellogg, it's hard to argue with Nicola. After all, I've had my own unsatisfactory encounters—the dangle from the chair and the subsequent diagnosis of an aging womb being just one among many—and it would be deeply gratifying to think that there might be some expert out there who sees, values, and listens to the real me, who is in tune with my womb and can soothe it into submission while honoring my deepest feminine truth. And maybe not bleeding uncontrollably through my clothes every month would be nice, too. So it is that I find myself squatting bare-vadged over a pot of tepid herbs on my bathroom floor, in search of my own little piece of womb wellness.

-

IT'S HALF PAST TWO ON A RAINY MONDAY AFTERNOON in July, and it hasn't been easy to find time alone in the house to steam my vagina. One or both of my children are almost always around during these school holidays, so arranging this sliver of soli-tude has been only slightly less challenging than the coordination of a major military exercise. I could be rewarding myself with a leisurely coffee—a proper one, brewed in the cafetière, with some of the good chocolate I hide where I know my daughters won't find it—or a lazy hour of Netflix. However, here I am in the name of research, naked from the waist down once again, squatting on the bathroom floor over a steaming pot of herbs I've ordered online. My undercarriage is get-ting warmer but everything else is getting cold; I've had to open the window as wide as it goes, because the residual pong of these leaves and petals will be hard to explain to a husband who's cynical about any cure that isn't dispensed by a certified professional in a white coat, or to daughters who would rather die than think about their mother's genitalia. I feel like a teenager smoking a sneaky spliff; my own teen-agers would be deeply and permanently embarrassed if they knew why the bathroom smells like a haystack.

If what I'm doing sounds unusual, it's because vaginal steaming—or vulval steaming, or V-steaming, or yoni steaming, as it's variously called by its practitioners—isn't quite as mainstream yet as getting a mani-pedi, but it's not far away. Catapulted firmly into the limelight by none other than Gwyneth herself when she wrote about it in 2015,[58] the practice has roots in many Indigenous cultures around the world. There is anecdotal evidence of the use of vulval steaming to promote gynecological health, sexual attractiveness, and postnatal healing on virtually every continent, from the Q'eqchi' Mayan people of Belize to the South African KwaZulu-Natal. Known variously in Korea as Chai-yok, in Central America as Bajo, and in Indonesia as Ganggang, steaming offers the user a ritual to engage almost every sense; fragrant herbs such as mugwort, wormwood, sage, thyme, and lavender are concocted by an elder or healer and steeped in near-boiling water. Sitting or squatting over this potent mix gives the user permission to take a moment for what's now known in millennial parlance as #selfcare. "It is an energetic release—not just a steam douche—that balances female hormone levels," Paltrow wrote after her visit to a local spa. "If you're in LA, you have to do it."[59]

As with so many womb wellness trends, from jade eggs to vagina-scented candles, where Paltrow leads, the world soon follows. V-steaming is big business, and not just in LA: enterprising female entrepreneurs have turned it into its own cottage industry, often with a strong selling presence on social media. Goddess Detox (491,000 Instagram followers at last count) sells steaming herbs and accessories along with its star product, "Pu$$y Power vaginal wash," and tells shoppers to "expect warm loving energy to enter your vagina and uterus."[60] Another online seller, Femmagic, endorses its range with the strapline, "Femcare for a Goddess."[61] Numerous other companies advertise steaming products whose names play freely on this theme of the female divine: plastic "thrones" that look suspiciously like potties, and voluminous, jewel-toned polyester "gowns" to be worn when steaming, either alone or with friends. There is, it should be said, a practical element to some of this merchandise; Goddess Detox says that with its gowns, "You can have yoni steaming parties with your friends without having to see each other's coota cats."[62] Steaming is pitched as a quintessentially fulfilling act: with the right products (one is led to believe) it can be both a group bonding ritual and an act of discreet self-care.

The English website from which I purchased my own herbs illustrates classic British reserve, with less Hollywood glamour and more soft-focus photographs of ferns, roots, and branches. No hyped-up promises here, just a gentle suggestion that "there has been [*sic*] numerous claims of herbal steaming helping with many areas of reproductive health. This includes helping regulate the menstrual cycle, reduce cramping and to help tone and strengthen the uterine lining."[63] Other sites promise the proverbial moon and stars to the woeful womb-owner, with claims that steaming can treat infertility, fibroids, infection, and poor libido; one site goes as far as suggesting that a V-steam regime can aid sexual abuse recovery, cleanse residue in the womb, and even minimize "labial gapping." As a midwife, I'm pretty sure that labia are meant to have at least a small gap between them (and I'm not sure if or why a bigger gap would even be a problem), but I can't deny the appeal of the other alleged benefits. One website invites me to "quiet your mind and reconnect with your yoni," something which, admittedly, I have never been able to do in that other hallowed arena of uterine healing, the gynecology outpatients clinic.

Squatting over my own homespun steaming setup, I'm not sure I'm exactly "quieting and reconnecting"; or at least, not in the way my inner goddess desires. The instructions that accompanied my herbs advised me to use a pot, but which pot, and how big? The saucepan I use for poaching eggs, or (the one I've gone for in the end) the bigger vessel I use for family-sized portions of spaghetti Bolognese? The sensation is not unlike someone blowing soft, warm breath onto my vulva: slightly creepy, but not unpleasant. It doesn't feel as if any of the steam is actually ascending my genital tract (perhaps my labial gapping is inadequate?) and by the time my ten minutes are up, the mixture has cooled so much that I can only feel the very vaguest whisper of heat against my thighs. The session hasn't left me with serious burns, as some online naysayers have warned, but as I stand and stretch my aching legs, I still have concerns.

In addition to the scent of my mother-in-law's potpourri, there's an undeniable whiff of cultural appropriation around the whole V-steam phenomenon. As with many other alternative therapies promising to restore health and balance to one's reproductive system, the vagina is poorly differentiated (if at all) from the uterus in many descriptions of vulval steaming, with the whole area identified simply as "yoni." The word itself is Sanskrit, although very few of the practitioners using this language and promoting "yoni health" in the West are actually of Indian origin. Some women may prefer to use the term "yoni" as an alternative to the clinical language of traditionally male-centered Western medicine, repurposing language and reclaiming anatomy at the same time, but the discourse around "yoni steaming" (and many womb wellness therapies in general) draws uncomfortably from a kind of exoticization of foreign languages and practices. Avni Trivedi, a London-based doula and osteopath, says, "I'm of Indian origin and 'yoni' isn't a word I grew up with, so it feels odd to have a version of my culture handed back in a different form. There's an element of exotification which feels inappropriate."[64]

Other critics of vulval steaming argue that the practice—and the industry now espousing it—is entrenched in misogynistic values, and that many of the cultures in which steaming originated actually used the ritual as a way of tightening the vagina to make it more sexually satisfying for penetrative male partners. "Steaming is a literal tool of the patriarchy,"[65] argues Dr. Jen Gunter, a Canadian gynecologist whose blogs, podcasts, and books follow the singular mission of debunking harmful myths about women's health. Gunter argues that the womb wellness industry claims to offer a cure for what she calls "vaginal mayhem":[66] the misguided belief that the vagina and surrounding structures are the seat not of divinity, but of chaotic toxins and dangerously deranged hormones that must be steamed, douched, fragranced, flushed, and "balanced" into submission. Merchandise is promoted under the premise that female genitalia must look, smell, or taste a certain way to be healthy or, more accurately, to be appealing and acceptable to the opposite sex. Gunter believes that any business taking advantage of women's shame by marketing unproven and potentially harmful products or treatments is "so predatory . . . That's taking patriarchy, wrapping it up in a pink bow, and saying it's feminism."[67]

On a practical note, there has been concern that steaming may cause real physical harm. One study from the University of Michigan at Ann Arbor found that vulval and vaginal tissues are uniquely permeable to volatile compounds that may be dangerous when released into the bloodstream.[68] Another meta-analysis of vaginal douching (not steaming, but a similar practice in which soaps or cleansers are inserted directly into the vagina) was correlated with an increased risk of pelvic inflammatory disease, ectopic pregnancy, and cervical cancer.[69] Other critics point to the simple fact that steaming the vulval area without proper care could result in superficial burns, and they argue that the creation of a hot, moist genital environment could create ideal conditions for the growth of harmful bacteria and fungal infections.

Perhaps most compellingly, some critics say that vulval steaming is emblematic of a psychologically harmful social imperative in which women are expected to engage in a constant, never-ending process of self-improvement. In their essay "Basically, It's Sorcery for Your Vagina: Unpacking Western Representations of Vaginal Steaming," psychologists Tycho Vandenburg and Virginia Braun situate steaming within a broader ideology of "healthism." The authors explain healthism as "the pursuit of not just of health [*sic*], but health optimisation, as a moral obligation . . . The subject seeks out, assesses and participates in, self-improvement strategies; without work, the body/ self is incomplete, but the task is endless."[70] Examining ninety online sources of information about vaginal steaming, the authors identified emerging themes of the female body being seen as essentially deteriorating and dirty, with vaginal steaming presented as a tool of optimizing self and life. Regardless of whether the women in the source material viewed steaming as a method of purifying a defective body or of pampering a body worthy of worship, in either model, the body itself is not "enough."

Indeed, if we strip away the thrones and gowns—the "royal" accouterments of steaming in its current ritualistic form—we are left with a truly exhausting idea: that the task of grooming, purifying, and sexually enhancing one's genitalia must always be attempted, but can never be completed. Like Sisyphus pushing his stone endlessly up a hill, only for it to keep rolling back against him, women in virtually every culture—whether industrialized or Indigenous, Western or Eastern—must contend with the thankless, endless, lifelong chore of modifying and maintaining their bodies, and by extension, their wombs. This chore is, as Vandenburg and Braun argue, "voluntary, but also obligatory," and ends only with the death of the body itself.[71]

Can the truth really be so bleak, and can thousands of happy V-steamers be wrong? After all, for every angry online doc shouting "Fraud!" and waving the red flag of quackery, there are hundreds more glowing reviews left by satisfied customers. The internet is bursting with testimonials from women who are delighted with their steams, teas, cleanses, supplements, coaching, and communities, and thanking the purveyors of womb wellness for improving their quality of life. For every woman claiming that a course of supplements helped her conceive, there is yet another claiming that vaginal steaming "keeps my man coming back for more," and another insisting that a course of expensive supplements has made her periods pain-free. To these consumers, womb wellness is neither the Sisyphean task described by Vandenburg and Braun, nor the toxic manifestation of vaginal mayhem condemned by Gunter. Rather, it is a satisfying and empowering endeavor with real and lasting results.

It is possible that these angry critiques of vulval steaming are inherently flawed, coming as they almost always do from a white-centric, Western-centric point of view that may be ideologically biased against less culturally mainstream experiences. This epistemological framing situates Western medicine as the authority on women's wombs and their needs; in return, women who promote and enjoy V-steaming are portrayed as "dupes" with little understanding of their own bodies. This narrative fits the model of what British philosopher Miranda Fricker calls "testimonial injustice." According to Fricker, "Testimonial injustice occurs when prejudice causes a hearer to give a deflated level of credibility to a speaker's word . . . The basic idea is that a speaker suffers a testimonial injustice if prejudice on the hearer's part causes him to give the speaker less credibility than he would otherwise have given."[72]

Women who practice vaginal steaming and other forms of womb wellness are often the very same women who feel poorly served by mainstream medicine; some may have been let down by healthcare providers in the past, others may have felt stigmatized or vulnerable, and still others may have found the financial cost of service engagement prohibitive. It is no wonder, then, that the most vociferous opponents of womb wellness are doctors. No matter how well informed and intentioned their arguments may be, it is not impossible that their prejudices against these "rogue" women have influenced—subtly or overtly—their opinions about unproven but popular alternative therapies. Critiques of womb wellness may carry an element of Fricker's testimonial injustice; dismissal of any perceived clinical or emotional benefit may simply be replicating the problematic dynamic of the patient–provider relationship.

Nicola Goodall agrees that she hears this narrative of prejudice and disbelief from many of her clients; she says that women who tell their doctor how they have used alternative therapies to cope with gynecological problems are met with alarming levels of vitriol. Nicola sums up the typical response as, "Do you think you are healing yourself? I'm in charge." Perhaps, then, the public discourse around vulval steaming is about more than outlandishly named products and dubious health claims; it's another manifestation of the endless power struggle between women and the medical establishment, a battle between lived experiences and desires, and the "right" kind of knowledge. Perhaps the gowns and thrones are not frivolous fripperies, but armor against a system built on exclusive knowledge of and power over women's wombs. When poet and activist Audre Lorde said, "Caring for myself is not self-indulgence. It is self-preservation, and that is an act of political warfare,"[73] she may not have had V-steaming in mind, but the comparison is apt.

–

AS FOR ME, THERE'S NO WAY OF KNOWING IF MY STEAM-ing session has worked, really, until my period arrives with its usual medieval brutality a few weeks later, a torrent of blood and cramp that has me rocking in the fetal position until my ibuprofen kicks in. In spite of my professional skepticism, I really wanted it to work; I wanted my aging, calcified womb to calm and soften. Of course, I may not have done my steaming "right"—one ten-minute session of hovering tentatively over a saucepan may be the wellness equivalent of taking half a paracetamol and expecting your brain tumor to melt away. But this is the problem and the paradox of womb wellness—as the industry is unregulated and its efficacy unproven, one can never be quite sure one is doing it right unless one believes.

This was not my first time at the womb wellness rodeo, having previously tried acupuncture (which was relaxing but didn't help me conceive), a cannabis-infused tampon (which made me feel so ill I had to whip it out in a panic mere minutes after insertion), and endless jars of supplements—fish oil, marine kelp, evening primrose, and a particularly vile wheatgrass tablet that had no discernible effect apart from inducing what we in Scotland call "the boak." These days, I grin and bear the monthly ordeal. I ponder the treatments offered to me by mainstream medicine—hormonal contraception to thin the lining of my womb, or removal of my uterus altogether—and I veer with increasing conviction toward hysterectomy until my cycle becomes more temperate on Day 3 or 4 and I resign myself to another month of the status quo.

Why did I persist? I haven't been trying to channel Hygieia, or commune with some inner goddess, but—like Michelle Hopewell and so many others—I've been driven to desperation by the severity of my symptoms and by the general inability of mainstream medicine to alleviate these issues in a way that feels individualized and empowering. It would be cheering to think that the womb wellness industry, with its impeccably groomed gurus and exclusive membership programs, could offer healing or help of substance. However, in spite of the supplements that bear the nebulous promise to "balance your hormones" and the pearls, pessaries, tablets, and teas that aim to "cleanse your womb," very few of these products or the people who promote them can provide definitive clinical evidence of measurable benefit. Womb wellness may offer a seductive vision of comfort and control, but when the industry and its wider offshoots report trillion-dollar profits without the data to back it up, the only thing getting detoxed may, in many cases, be the consumer's bank account.

If womb wellness is to have any chance of reducing harm, rather than causing it, then it needs to be offered by reputable sources with safe, proven, demonstrable results. If and when those results are achieved, they should be available to all at a reasonable cost, without having to pay out the better part of a year's wages or subscribe to an endless "program." Nicola Goodall has seen numerous so-called masters of womb wellness come and go during her career, often with few credentials to justify their extortionate fees.

"This idea that you just ask somebody who knows something about it, and then you say, 'I'm the master teacher, let's make thousands of pounds'—that's the dangerous thing. I see all sorts of dangerous shit like that in the womb industry."

When I ask her about one practitioner who charges £10,000 for six months of personalized cycle coaching and womb optimization, Nicola balks: "Ten grand? What about a woman who lives down the road in the flats, who can't even feed herself? Looking after your womb is not a luxury. It's part of daily life, so we really need to make sure that people know what they're doing."

Until this affordability and professional accountability are embedded in the womb wellness industry, what began as a reaction against the predominantly white, male, moneyed world of obstetrics and gynecology may yet bear some of the worst qualities of the system it seeks to oppose.

Menopause

ENDINGS AND BEGINNINGS

We turn not older with years, but newer every day.

—EMILY DICKINSON, LETTER TO LOUISE NORCROSS

The womb announces itself month after month for thirty to forty years, each cramp and tender breast and plummeting mood and sudden, scarlet streak of blood saying hello again and again and again, whether welcome or unbidden—a sigh of relief or a sign of loss—until gradually, that monthly greeting becomes less frequent and more erratic. Finally, it lapses into silence. This, then, is menopause: a condition that can only be diagnosed retrospectively, as it describes a year's passing since the date of a final period. The average age of menopause is roughly fifty-one to fifty-two years, although premature cessation of menstruation can sometimes occur much earlier, and some people don't experience their last period until their late fifties.

Nowadays, symptoms of menopause and perimenopause—the time of hormonal flux leading up to the final menses—get a lot more press than the menopausal uterus itself. Perhaps this focus is apt, as this stage of reproductive life is initiated not by the uterus, but by the ovaries' diminishing production of estrogen and progesterone. For an average of seven years before the last period, and for many years thereafter, this change in ovarian function can cause a wide range of emotional and physiological changes, from anxiety, low mood, irritability, and "brain fog" to hot flushes, fatigue, night sweats, palpitations, and achy joints. Declining estrogen levels can cause thinning and dryness of the vaginal tissues, a painful and once taboo condition that is now catered to by a burgeoning industry of topical creams and lubricants. What was once *verboten* is now the hot topic in the media; these days, a new menopause manual seems to hit the bookshop shelves each week, and celebrities hoping to tell their own tales of "my menopause hell" jostle for space on daytime TV couches.

As for the uterus itself, what—if anything—does it actually do in menopause apart from . . . very little? No longer prompted by monthly hormone fluctuations to thicken and shed its lining, the menopausal womb decreases in size and thickness by roughly 20 to 30 percent. Just as many menopausal people find their shape changing during these years—estrogen-fueled curves and collagen giving way to a gradual sag and ebb—the size ratio of the main body of the womb to the cervix changes, too. In some cases, a general weakening of the uterus and its surrounding structures can cause the organ to fall out of its usual position, securely supported by the sling-like ligaments and muscles of the pelvic floor. This kind of prolapse can be either partial, with the womb bulging down into the top of the vaginal passage, or complete, with the uterus seen and felt at the vaginal entrance.

Like so many aspects of reproductive health, menopause means different things and manifests in myriad ways for different people, with some experiencing almost all of these troublesome symptoms and some having few or none; and like other ages and stages, this normal physiological phase of a woman's life was nameless, poorly understood, and highly stigmatized for much of modern history. Aristotle and Soranus both wrote of the apparent link between the end of menstruation and the loss of fertility, noting that this could occur anywhere from fifty to sixty years of age. Medieval German abbess and mystic Hildegard von Bingen provided one of the first-known female-written descriptions of menopause in her text *Causae et Curae* (Causes and Cures); she wrote that "the menses ceases from the fiftieth year and sometimes in certain ones in the sixtieth when the uterus begins to be enfolded and contracts so that they are no longer able to conceive."[1] This gentle image of the womb folding and contracting in on itself hints at a period of inward-turning and introspection—a nuanced description from an author who, age fifty-two at the time of writing, may even have been reflecting on her own experience of menopause.

Over the years, this phase of female existence acquired increasingly derogatory nicknames—among them "women's hell" and "death of sex"—until, as happened so often in the nineteenth century, the phenomenon was renamed by a man. French physician Charles de Gardanne coined the term "la ménespausie" in his 1816 book *Avis aux femmes qui entrent dans l'age critique* ("Advice to women entering the critical age"),[2] describing this time as a pathological period beset with dangers and ailments, and later shortening the term to the familiar "menopause" in a subsequent treatise on the subject in 1821.[3] In his repeated assertion that women are ignorant of their own condition, de Gardanne set the tone for a narrative of menopause that continued with few exceptions over the following decades. Just as Gooch and his contemporaries insisted that "irritable uterus" was a psychosomatic disease that could prey on women's inherent weakness, so nineteenth- and early twentieth-century physicians agreed that menopause and its attendant symptoms could be brought on by stress, bad news, and overwork, including duress from "unwomanly" occupations such as fishmongery. As one might expect, psychotherapist and notorious pathologizer of women Sigmund Freud picked up this notion of the neurotic menopausal female and ran with it, claiming in 1913 that these poor, unfortunate creatures who had ceased to menstruate soon began to exhibit "typically sadistic and anal-erotic traits which they did not possess earlier during their period of womanliness."[4]

Menopause continued to gain a bad rap well into the twentieth century, reaching a fever pitch in the 1950s and '60s, a time when dominant Western thought pushed back hard against emerging ideas of an empowered postwar femininity. Women's liberation and mass bra-burnings may have been just around the corner, but mainstream culture still clung hard and fast to traditional gender roles and qualities. In this framework, menopause was a tragedy: a loss of femininity to be mourned, and a decline to warrant urgent concern. Helene Deutsch, a Polish American psychotherapist who had studied under Freud in Vienna prior to her emigration, wrote in 1958 that a woman's "service to her species ceases . . . with the lapse of her reproductive service, her beauty vanished and usually, the warm, vital flow of the feminine emotional life as well."[5] In his 1966 book *Feminine Forever*, American gynecologist Robert Wilson reframed menopause as a perilous time of hormonal deficiency, writing, "No woman can be sure of escaping the horror of this living decay," and advocating estrogen replacement therapy as the only possible cure for the "extreme suffering and incapacity" of a withered womb and its sexually withered owner.[6]

Advertising campaigns of that era made it very clear that menopausal "suffering" and "horror" posed the biggest threat not to women, but to their husbands. One 1960s advert features a photo of a disgruntled bus driver with the caption, "He is suffering from estrogen deficiency," next to a picture of a middle-aged female passenger, her face scowling and accusatory, and the punchline, "She is the reason why."[7] Premarin, the first widely available form of hormone replacement therapy, appears to have been marketed as a kind of marriage-saving miracle cure, guaranteed to prevent one's wife from metamorphosing into the archetypal haggard harridan. One Premarin ad from 1966 features an image of a slim, attractive woman talking to two men at a party; she appears to be laughing enthusiastically at something one of the men has said, and the caption suggests simply, "Help keep her this way."[8] The subtext is clear: she may be old on the inside, but on the outside, she remains as young, appealing, and attentive as ever. Even doctors were encouraged to play their role in maintaining marital bliss: "The physician who puts a woman on Premarin," states one ad, "usually makes her pleasant to live with once again."[9] There is no mention of how the woman feels, or what might be "pleasant" for her; the effect on her partner is of sole importance.

Thankfully, science and society have moved on since the 1960s; research into HRT continues to shed light on the safest, most effective forms of treatment, and the discourse around menopause now focuses almost exclusively on the well-being of the person going through it. However, the newest ideas and best points of practice have not always filtered through to those who should be most knowledgeable about this transformative stage of women's lives. In many ways, the medical establishment's confusion around menopause and the best way to treat it has continued until the present day—so much so, that even the best-informed female doctors can sometimes struggle to identify and manage their own experiences.

Dr. Zoe Hodson, a GP in Manchester, says her special professional interest in menopause arose after a time of personal challenge: "It was the usual thing: I was studying [perimenopause] day and night and didn't realize I was going through it. It came and it floored me, and I actually had to have time off work, which I haven't done in twenty years." Realizing that she and many of her colleagues had no idea how to treat menopause holistically and effectively, Zoe reflects, "Women don't know; healthcare professionals don't know. [As a GP] you sort of know how hormones behave, but still all these menopausal women are coming in, and I started thinking, I'm just not very good at this. And it's really frustrating—everywhere I tried to get information from, it still didn't feel right. It just felt as though it hadn't been updated or refreshed."

Although Zoe has since developed her own focus on menopause and now trains other doctors to have a deeper understanding of this stage of life, she candidly admits that the subject is still shrouded in confusion and misinformation. Add to that a persistent social stigma around declining sexuality, compounded by an often unexpected feeling of bereavement at the final loss of fertility, and it's no wonder that the menopausal years continue to challenge even the most self-aware woman (and her doctor).

Recently, new voices have entered this discourse around menopause, suggesting that rather than treating this phase as pathological and problematic, women should celebrate their liberation from the tyranny of the uterus and its monthly handmaidens, pain and bleeding. Books with bold, confident titles, such as *Perimenopause Power*, *The Menopause Manifesto*, and *Still Hot!* evoke a spirit of empowerment, confidence, and renewed sexuality. Perhaps one of the most emphatic declarations of postmenopausal freedom comes from Belinda, a character in the BBC series *Fleabag*, created and written by Phoebe Waller-Bridge. On seeing thirty-two-year-old "Fleabag" brooding over her cocktail after an awards ceremony for women in business, fifty-eight-year-old Belinda offers some words of menopausal wisdom:

"[Women] have pain on a cycle for years and years," she explains, adding that "we carry it within ourselves." In contrast, Belinda promises, menopause is "the most wonderful fucking thing in the world" because at that point, "you're free. No longer a slave, no longer a machine with parts. You're just a person."[10]

If, as Belinda says, women are born "with pain built in," and the site of that pain is so often the uterus, then it makes perfect sense that the organ's final silence can also be a woman's clarion call to joy. The womb's retreat offers freedom not just from the physical pain of menstruation, birth, and myriad illnesses such as fibroids and endometriosis, but also freedom from the onerous burden of being a woman in a society that prioritizes sexual availability and fecundity above all else. "You're just a person," says Belinda of the postmenopausal woman: not a slave to hormones, nor a potential reproductive mate, nor a gestatory vessel, but just a person, finally able to fulfill her own humanity.

"If we could move toward that," says Zoe Hodson, "and accept women as the brilliant, functional, fabulous beings that they are, then it would make this transition easier."

Much has been written about the empowering potential of menopause. The dominant contemporary narrative exhorts women to embrace this stage of life as an era of dynamic, transformative potential—a time to shrug off the responsibilities of child-rearing or career-building, perhaps, and to explore and indulge one's own desires. It feels apt, here, to remember the words of an author who posits menopause as a kind of metaphorical womb in and of itself: a time and space in which a woman must birth the fullest, truest version of herself; a kind of golden hour of self-actualization between the sexual and reproductive demands of young womanhood and the final transition of death. In her essay "The Space Crone," Ursula Le Guin—a writer of acclaimed and highly imaginative science fiction—imagines the menopausal woman as a magical, liminal form, heavy with her own gestation. Le Guin suggests that after fertility has waned, a woman who still wishes to live a full and satisfying life "must become pregnant with herself, at last. She must bear herself, her third self, her old age, with travail and alone . . . That pregnancy is long, that labour is hard. Only one is harder, and that's the final one, the one that men also must suffer and perform."[11]

Admittedly, this kind of self-actualization is a privilege only available to those with the kind of resources and socially equitable environment that make it possible to survive and thrive at any stage. However, regardless of the obstacles to becoming a truly fulfilled Space Crone, the metaphor of menopause as womb feels more powerful and authentic than the tropes that have preceded it: neither a horror show of disease and decay, nor a Pollyanna-ish picture of unfettered girl power. The Space Crone, "pregnant with herself," at last embodies the light and shade of the female experience. Like the birth of a child, the birth of a woman in menopause is ugly, beautiful, dangerous, and miraculous, a transition both corporeal and spiritual from one world to the next.

Hysterectomy

ABSENCE AND TRANSITION

They did it finally! Dug it out
With their stainless steel and gloved hands
A mass of jellied substance
To be divided equally into sterilized masons.

—PHYLLIS DOYLE PEPE, "HYSTERECTOMY"

Many women part with their womb well before the long, slow goodbye of physiological perimenopause has begun. Some older women, too, find themselves plagued by uterine woes years after the final menses has come and gone. There are some problems that simply cannot be solved by medication or careful management, either traditional or alternative. Some cancers are not safe to treat by pharmacological means only; some postpartum hemorrhages cannot be staunched by drugs; some ostensibly benign conditions, like prolapse, heavy periods, or persistent fibroids, can decimate a person's quality of life. On these occasions, the only solution is hysterectomy: removal of the uterus itself.

After Cesarean section, hysterectomy is the second most common surgical procedure performed on women of childbearing age, with over one million performed worldwide each year.[1] In America, one third of women will have had a hysterectomy by the age of sixty:[2] an army of the walking, wombless wounded. Today's patients have the benefit of hindsight and history. While not without its potential complications, hysterectomy is generally much safer than it was at the time of its origin. In 120 AD, Soranus of Ephesus is said to have treated a gangrenous prolapsed uterus by removing the offending organ via the vagina. The patient—like many more over the following centuries—did not survive. In 1670, English man-midwife Percival Willughby recorded one of the most detailed early accounts of a hysterectomy resulting in a live patient—all the more remarkable for the fact that the procedure was performed by the patient herself. Local woman Faith Raworth was driven to desperate measures after an attempt to carry "a heavy coale" left her with a persistent uterine prolapse. Having tried and failed to tuck her womb back into its rightful position ("Shee oft put it up, but it would presently fall down againe"), Faith decided to take matters into her own hands.

Willughby writes:

> *Being troubled, and discontented, and wearied with this affliction, in hopes to cure her self, shee went into the garden, and, laying hold on it [her uterus], drew it, and cut it forth.*

Perhaps unsurprisingly, "a great flux of blood followed," along with the unpleasant realization that Faith had also severed her bladder and part of her vagina. Willughby was summoned, and fashioned a temporary repair "with double twisted silk," but the stitches soon separated and Faith was left permanently incontinent. He recalls:

> *Her urine issued again by the old breache. Shee lived several yeares with this affliction, and died uncured, her water always comming night and day, insensibly dribbling from her.*[3]

While this tale is often told by medical historians with an undertone of wry mockery (Silly woman! What did she expect?), Faith Raworth's story is, in fact, a sad testament to the lengths to which a woman will go to manage her own gynecological health, regardless of the painful and sometimes unthinkably grim consequences.

Hysterectomy remained problematic for both patient and practitioner until well into the nineteenth century, when medics in both England and America pioneered the abdominal hysterectomy (in which the womb is removed via an incision in the abdomen, rather than the riskier vaginal extraction) with varying degrees of success. In 1885, Edinburgh physician Thomas Keith bemoaned the efforts of surgical colleagues in London, Berlin, and Washington, some of whose patients suffered mortality rates as high as one in three: "So far as hysterectomy has thus gone, it has done more harm than good, and it would have been better that it had never been. If these be the best results that surgery can give, then the sooner this operation is laid aside the better."[4] Fortunately, Keith had the privilege of practicing during Scotland's golden age of antisepsis, and his introduction of new techniques to cauterize the cervix during hysterectomy brought his patients' mortality rate down to about 8 percent. Thanks to other innovations in infection control and anesthesia, hysterectomy was no longer a last-resort option offered only to the moribund. This, no doubt, would have been welcome news to the women who would otherwise have been offered popular treatments of the time such as leeches between the labia (for prolapse) or carbolic irrigation of the uterus (for heavy bleeding).

Over time, hysterectomy became a common tool of the twentieth-century gynecologists' trade, with post-op recovery being slow but generally safe enough for the procedure to be brought into standard practice. There was little change to the operation itself until 1988, when the first laparoscopic or "keyhole" hysterectomy was performed by a Pennsylvania doctor, and again in 2002, when a Texas team carried out the first robot-assisted hysterectomies, in which remotely controlled mechanical "arms" plumbed the depth of the patient's pelvis. With minimally invasive procedures now the norm, hysterectomy has come a long way since Soranus took up his knife and Faith Raworth took off her own uterus.

For women who undergo hysterectomy, the experience can evoke a broad spectrum of emotion, from bereavement at a perceived loss of femininity to joy at the liberation from pain and bleeding. Yvonne, a sixty-four-year-old nurse from Kent, England, knew that hysterectomy was a necessity following a diagnosis of endometrial cancer. Nevertheless, her relief at being offered lifesaving treatment was tempered by sorrow, and even a sense of betrayal.

"I felt sad," she wrote to me in a long, poignantly detailed email, "that this uterus that had given me my beautiful family was now turning against me, trying to get me, secretly growing another life form, but one that was unwanted and would be a danger to me." There was grief, too, she says: "Although I knew I did not need my uterus any more, I was feeling a sense of loss about losing an organ that is essential to being a woman." After the surgery, Yvonne's doctor showed her some images that brought this transition into stark relief: "The registrar showed me before-and-after pictures of the site of my womb. The picture of my womb in place was amazing, so textbook, with the fallopian tubes splaying out into my pelvis. The colors were amazing, and I could not believe how perfect it all looked," she recalls. "The next picture showed this empty space and, at the time, I remember feeling a huge sense of loss and relief. Really mixed emotions." In spite of this ambivalence, Yvonne is proud that it was her own knowledge of her body and her dogged pursuit of treatment that led to an accurate diagnosis and successful surgery: "Since that time I've wanted to let women know—listen to your body and don't ignore any out-of-the-ordinary discharge or bleeding. I'm pleased," she says, "to be sharing this story with you."[5]

Many other women contacted me to share their hysterectomy stories, too, even—and sometimes, especially—if their own surgeries were life-enhancing rather than lifesaving. Denyse, a retired teacher from Australia, describes herself as "evangelical" about hysterectomy after the procedure freed her from a lifetime of debilitating pain and bleeding. From her first period at the age of twelve, Denyse suffered a monthly ordeal that, in her words, "not much could stop." Painkillers and prescription medications were ineffective, and she found herself having to take days off school and work. Finally, when Denyse was thirty-eight, a doctor recommended hysterectomy after an ultrasound revealed large fibroids in her uterus.

"I was on board with that!" says Denyse, who had had two children by that time and felt like her family was complete. After an initially slow recovery, Denyse describes herself as "free," advancing rapidly through her professional career over the next ten years "without worrying and wondering about periods." She's now keen to share her experience with others, and she does so often in online forums and social media.

"We tend (then more than now) not to talk about it," says Denyse, but without more awareness, "how would others know of this release from being tied to the house or bed by blood and pain?"[6]

One woman's tale of liberation, though, stood out from the rest: bittersweet, tragic and, ultimately, redemptive. "I actually became a gynecology nurse to help protect and respect women, their uteruses, and their vaginas," writes Stephanie from New York. "I myself had a hysterectomy aged forty-five for a uterine prolapse," she says, "and I was relieved to have the hysterectomy because of the discomfort caused by my prolapse, but also for another reason: I was raped as a teenager.

I had my period at the time, and was left having to clean up a lot of blood afterward. I don't think there was a monthly period from that night onward that wasn't a trigger to remember the rape, so removal of my uterus freed me from that monthly image. It also meant no more smears, which had been an emotional challenge for me."

Stephanie is keen to point out that, awful as her story may be, hers is likely to be a much more common experience than one might think. "I would suspect that I'm not the only woman to whom this has happened," she says, "but it's something very difficult to verbalize, and it's a taboo hidden within the taboo of rape. There will, of course, be many women who had their period when assaulted."[7] While Stephanie's hysterectomy was performed primarily for medical reasons, for her—and possibly for untold others—the procedure had the welcome side effect of release from a monthly reminder of trauma.

For Stephanie, Denyse, Yvonne, and many women like them, the uterus is a nexus of pain, and its removal brings autonomy and peace. No story of hysterectomy is complete, though, without acknowledging the countless other women for whom the procedure symbolizes a premature and unwelcome end to fertility. Natalya, a solicitor from Leeds, held on to her womb for years, hoping for a family in spite of often debilitating pain and bleeding caused by severe endometriosis. Having had ten laparoscopic surgeries to treat the lesions scattered throughout her pelvis, Natalya persevered with invasive, costly fertility treatments—first intrauterine insemination, then IVF—"Still harboring a hope that it just might happen," she says, "but unfortunately, it never did, and my husband and I split up."

After years of worsening pain, Natalya's situation became unbearable: "Once I hit my late thirties, I was having a period every other week and in absolute agony. The endo was now on my bowel, bladder, and ovaries. I managed the pain for several years with various drugs, but it got to a stage where it was affecting my day-to-day life."

Finally, a new consultant performed a final laparoscopy, which revealed the extent of Natalya's disease: "My ovaries had disintegrated, my bowel and bladder had attached to each other, and my womb was just a complete mess. This was one of the worst cases my consultant had seen: I needed a full hysterectomy and needed one quite quickly. There were never going to be any children. I was devastated: I didn't feel like a woman. I let myself down."[8]

In retrospect, Natalya is grateful that the procedure restored her quality of life—"I was instantly pain-free, and physically it was the best thing for me"—but she feels a deep and lasting sense of loss for her uterus and the family it never gave her, and admits to being unexpectedly upset by sharing her story. "It has affected me quite deeply," she says. "In fact, I didn't think it would bother me to talk about it. I'm quite surprised how I feel."

Natalya may be surprised by the depth of her emotions around hysterectomy, but she's not alone. Andréa Becker, a sociologist researching hysterectomy experiences, says that many women mourn their femininity for years after the operation.

"One woman I spoke to has a sticky note on her mirror that says, 'You're still a woman.'" A reminder to that woman—and to all of us—that the uterus is often inextricably linked to a person's identity, and that its removal is as powerful as its presence.

Indeed, evidence suggests that hysterectomy can have a substantial effect on a woman's mental health. In a study that followed over 2,100 women for twenty-two years, hysterectomy was associated with an increased rate of new diagnoses of long-term mental health issues; specifically, a 6.6 percent higher risk of depression and a 4.7 percent higher risk of anxiety. For those women who had a hysterectomy between the ages of eighteen and thirty-five—arguably, the prime years of fertility and childbearing—the risk of depression was 12 percent greater.[9] Overall, hysterectomy was associated with a range of disorders, from agitation and reduced psychosexual function to psychosis. The NHS Online makes a brief, one-line acknowledgment of these potentially life-changing effects: "Having your uterus removed," says the website, "can cause you to have feelings of loss or sadness."[10]

Womb or no womb, many women might rightfully bridle at the suggestion that one's identity begins and ends with one's reproductive potential. The idea that a person without children should resign herself to a sad, lonely life of gazing wistfully at self-penned sticky notes and longing for the babies she never had is dangerously retrograde, and should be resigned to the waste bin of history along with every other harmful cliché that puts female identity into a tidily reductive box. As Denyse, Yvonne, and many others like them will testify, hysterectomy can broaden—not narrow—one's horizons under the right circumstances. However, what they may not know—and what was as much of a shock to me as it may now be to you—is that scientists are only just starting to understand how fundamentally hysterectomy can affect the very seat of identity, emotion, and day-to-day function: the brain.

TO JOURNEY TO THE CUTTING EDGE OF RESEARCH INTO the uterus-brain connection, we must travel to Tucson, Arizona, where, nestled between the cactus-studded Saguaro National Park and the snow-topped peaks of the Rincon Mountains, a team of behavioral neuroscientists have been doing something unusual with rats. True, the University of Arizona markets itself as a kind of desert wonderland, where "everywhere you look is filled with wonder" and your imagination "ignites" beneath a sky of "cotton candy sunsets and diamond-filled nights,"[11] but what the prof and her rats have revealed is more outlandish than even the hammiest prose.

Stephanie Koebele, the study's lead author, and her team prepared four groups of rats: one group had just their ovaries removed, one had just their uterus removed, one had both ovaries and uterus removed, and a control group had a "sham" operation (opening and closure of the abdomen with no actual removal of organs). Six weeks post-op, the rats were trained to navigate a maze, and the results were shocking enough to make Jean-Martin Charcot turn in his elegant Montmartre grave.

The rats who had had an operation to remove the uterus alone made more navigational mistakes and overall were more challenged by the maze. There was no other difference between them and the other rats—no known hormonal deficit or change, no special drugs, no other handicap—and yet the simple absence of the uterus was enough to have a demonstrable effect on these animals' cognitive abilities. If this organ (or lack thereof) has an influence—even a small one—on memory and spatial awareness, then what other aspects of thought and function might it also affect?

"The dogma is that the non-pregnant uterus is dormant," Koebele and her coauthors write. In other words, we've believed for millennia that a womb's purpose is solely reproductive. It menstruates to prepare for a baby, it gestates and expels a baby, and then it waits, silent and useless in its owner's pelvis, until death. However, the Arizona team argue that their findings point toward "an ovarian-uterus-brain system that becomes interrupted when the reproductive tract has been disrupted, leading to alterations in brain functioning."[12]

The nonpregnant uterus, they suggest, "*is not dormant.*"

The womb, the team believe, is communicating with the brain in a way that is fundamentally powerful, albeit poorly understood. There appears to be some kind of critical uterine-cognitive dialogue—a conversation that stops abruptly when the womb is removed and sent to the hospital incinerator, like hundreds of thousands of these organs are each year.

Some would argue that this new data is illuminating and empowering: the womb can do amazing things never before thought possible! Women are beautiful, mystical, multifaceted beings whose reproductive systems are far more sophisticated than science has yet understood! And yet, one could also regard these new findings with consternation, arguing that the Arizona study is yet another chapter in the reductive narrative that places a woman's brain at the mercy of her uterus, mischievous and malfunctional as that organ may be. Perhaps, though, this research opens a door to a less binary, more nuanced way of thinking about the relationship between behavior and reproductive biology. Instead of assuming that a potential womb-brain relationship is something primitive, dangerous, and shameful, it may be more productive—and ultimately, more accurate—to consider the many possible cognitive and emotional implications of such a connection. A growing body of recent evidence has given credence to a similar relationship between the brain and the gut, a system that remains relatively unburdened by the weight of sociosexual constructs.[13] In time, the womb may escape its stigma as the seat of womanly folly; like the gut, the uterus may yet be recognized as one of many complex factors which influence—and even enhance—a person's thoughts and feelings.

If anything, those lab rats should serve as reminders to practitioners and prospective patients that the effects of hysterectomy may not yet be fully understood, and as such, the procedure should always be approached with due diligence. A 2015 study from the University of Michigan adds further credence to the argument for caution: of 3,397 women undergoing hysterectomy for benign (i.e., non-life-threatening) conditions, 18 percent were found to have "unsupportive pathology" for the procedure.[14] In other words, the findings of the surgery did not support the previously determined need for hysterectomy.

Nearly one in five operations were, in essence, unnecessary. This stark finding is now reflected in statutory guidance; the American Congress of Obstetricians and Gynecologists, for example, recommends other forms of hormonal and medical treatment as "primary management" in benign cases before hysterectomy is considered or performed.[15] Dr. Christine Metz says that, by and large, "If you have a healthy uterus, you should keep it for as long as you can."

FOR SOME PEOPLE WITH A HEALTHY WOMB, THE SOCIAL and emotional benefits of hysterectomy far outweigh any potential physical or cognitive side effects. Ryan Sallans is one of these people, and his story begins in 1986 in Aurora, Nebraska. Young Ryan is seven years old at that time—a self-described "country kid" in this small town of less than three square miles and five thousand residents. These days, Aurora declares itself a place "where the possibilities are endless." In 1986, though, seven-year-old Ryan thinks he has a pretty good idea of what life has in store for him. He knows he likes climbing trees and being with animals. He knows he feels free and happy when he's on his own in nature, lost in the moment of play. There's just one problem: to his parents, and to everyone who knows him, he is not Ryan yet. In 1986, he is Kimberly Ann Sallans, and he is a girl. This, it suddenly occurs to this easygoing, towheaded child, is a problem.

"We lived on an acreage with a swimming pool in our backyard, and all these trees and gardens we maintained, and I was (and remain to this day) an outdoors person. I need nature," Ryan tells me during our video call. He is energetic and keen to share his experience in spite of the fact that it's still relatively early in Nebraska; his desk is tidy, his beard is neatly trimmed, and the memory of his free-range childhood is still fresh in his mind.

"I remember I came in from playing outside in the trees," he recalls, "and I was standing in one of our bathrooms washing my hands because we were about to have dinner. And I looked at myself in the mirror and a voice said to me, 'You're female. You're not a boy.' And I just had this moment of dread of—not to be offensive to people—but dread of a female body. This moment of dread consumed me, because I was still living in this fantasy that I hadn't really assigned a specific label to my body yet. And I remember hearing that voice and thinking to myself, 'This sucks.'"

This proved to be a formative moment of cognitive dissonance: a conflict between Ryan's sense of self and the identity projected onto him by the "voice" of society and some higher order.

"I just couldn't understand why God made me female," he says. "I was like, that's not who I am. So why do I have to live this way?" At first, this problem of femininity was vague, if undeniably troubling, but when puberty loomed like the dark vortex of a tornado on that flat midwestern horizon, womanhood became a very real and terrifying prospect.

"I remember in sixth grade, a bunch of us girls were in our library in our elementary school," Ryan recalls, "and they had a nurse or a midwife come in and she started showing us stuff about periods. So I was aware it was going to happen to me. And then in middle school, the girls would write notes in red ink and put them in people's lockers when they found out they had their periods. It wasn't meant to be mean—it was like some kind of menstrual club—but I was like, I hope I never get one of those. I do not want that. And then at around age twelve, I actually started menstruating and I was terrified." Ryan's horror at this monthly reminder of his biological identity was compounded by the fact that the periods themselves were heavy and painful: "Some people might have positive experiences or learn to love the menstrual cycle," he admits—one of the many times during our conversation when he makes a point of being as inclusive as possible—"But for me, it was never, ever that. I always had very, very heavy periods. It was just terrible, and embarrassing."

Month after month, the trauma of menstruation recurred. As time went by, Ryan's emerging awareness of his deeply felt identity proved painfully irreconcilable with his physical self. His struggle plunged him into depression and drove him toward increasingly self-destructive behavior, culminating with anorexia during his time at university. At first, Ryan says, he welcomed the fact that starvation had also stopped his periods.

"One of the things I actually thought was really cool was that I didn't have to worry about that monthly cycle anymore. I was like, oh, this is going to make my life easier. I really welcomed that side effect of amenorrhea."

Over time, though, Ryan's discomfort with his female anatomy gave way to a growing awareness of another option—a way through and out of the dissonance that had dogged him since that horrifying epiphany in the bathroom at age seven. Ryan discovered a book of photographic portraits of people who had been assigned female at birth and later transitioned, hormonally and/or surgically, to manhood. At the sight of these powerful images, Ryan realized that these were his people, and this was his path. He had "top surgery" [removal of breast tissue] first, and then he began to take testosterone, altering his body and mind at the most fundamental hormonal level. He dreamed of the day when his body would look on the outside the way it felt on the inside, and he knew intuitively from the start that this transformation would also entail the removal of that organ that made him the woman he didn't want to be—the small, clenched fist in his pelvis that pummeled him with pain and bleeding every month.

"I knew my ultimate goal was hysterectomy," says Ryan. "I started to experience more and more severe cramping, which is actually quite common for trans men on testosterone, and my cramping was getting to the place where three out of the four weeks of the month, it wouldn't stop, even at night. It was keeping me awake." Whereas some people—the #periodsoptional crew—use continuous oral contraception to stop menstruation and the pain that often accompanies it, most transitioning men find this option intolerable; the progesterone and/or estrogen in the Pill would perpetuate unacceptably "female" functions and features. For Ryan, there was only one way forward:

"I didn't want treatment, I didn't want any type of way to manage the pain. I just wanted all that stuff out."

As Ryan searched for an inclusive gynecologist and began to wrangle with the intricacies of American health insurance, he also found himself journeying through a system that, broadly speaking, is guided in many countries by standards of care set out by the World Professional Association for Transgender Health (WPATH). Published in 2012, "the standards," as many healthcare professionals call them, aim to create a consistent global pathway for patients and practitioners alike. In the case of hysterectomy for individuals transitioning from female to male, WPATH recommends that patients should have "persistent, well-documented gender dysphoria," full mental capacity to provide consent and make informed choices, age of majority, well-controlled medical or mental health concerns, if any are present, and twelve continuous months of hormone therapy prior to surgery.[16] While its parameters may be clear, this pathway can be circuitous and frustrating, often involving long waiting lists, lengthy evaluations by numerous caregivers and interagency collaboration between GPs/family physicians, specialist gender identity clinics, and hospital-based surgical/gynecological wards. It's impossible to know exactly how many men like Ryan navigate this process every year, largely because official data from healthcare providers, regulatory bodies, and insurance companies often fail to aggregate patients by gender or even, in some cases, to recognize or include transgender identities in any way. One 2019 review from a team at New York University estimates that 42 to 54 percent of transgender men have had gender confirming surgery, although this may not always include removal of the uterus or other female genitalia.[17] The 2015 US Transgender Survey, the largest of its kind with over 27,000 respondents from across the United States, reported that out of all self-identified trans men, a total of 71 percent had either already had a hysterectomy or wanted it "some day."[18]

Trans men's reasons for seeking hysterectomy are as diverse as the men themselves. Dr. David Gerber, consultant psychiatrist at the Sandyford Gender Identity Clinic in Glasgow, says that many of his patients report the kind of testosterone-exacerbated pain experienced by Ryan Sallans, as well as an aversion to periods in general and all that they represent.

"Many trans men hate the fact that they menstruate," David says. By extension, "the idea of being pregnant, for those trans men, is abhorrent." It's important to note at this point that many trans men do choose to retain the internal and/or external genitalia with which they were born, whether for the purposes of pregnancy or otherwise. While Thomas Beatie made waves around the world as the so-called first pregnant man when he and his bump appeared on the cover of *People* magazine in 2008, other sources indicate that trans men have used their wombs to gestate their own babies since the 1980s, if not before.

"It's not a male or female desire to want to have a child," Beatie told Oprah Winfrey in a subsequent interview. "It's a human desire," he said;[19] a desire that continues to be felt by many trans male birth parents around the world. (Once again, incomplete data collection obscures the true number of such parents, although Australia, for example, documented two hundred and fifty births gestated by trans men from 2009 to 2019.[20])

The sociologist Andréa Becker says that, in fact, many trans men attach neither positive nor negative emotional significance to the uterus; rather, they regard it with a kind of neutral detachment. To these men, the womb is merely a superfluous object to be removed, more like an appendix than the nexus of their sexual identity.

"I find neutrality a lot more pervasive among trans men," Andréa tells me, "where they're like, 'Well, this was just something I needed to do. Like, if I needed to get my tonsils out, that would just be another surgery. It wasn't supposed to be there in the first place.' Of course, there are other positive stories of 'This was great, I feel so much more myself,' but hysterectomy struck me as [often] like it could be an administrative task."

For Ryan, who had felt so strongly that removal of his uterus was always the "ultimate goal," the week before the surgery itself brought an unexpected deluge of conflicting emotions. His decision to get his ovaries removed, too, compounded this wave of ambivalence.

"After all the things I'd done," he recalls, "like chest surgery, testosterone, and flying over to Belgrade for my lower surgery, I had no regrets. I was cool and calm as a cucumber going in. But the week before the hysterectomy, I went through a huge state of grief and questioning, because I knew that I was getting rid of any way of being able to pass on my own genetics. That's a big decision at twenty-six years of age, so I really had to allow myself to have that grief and go through that."

Now, nearly twenty years later, Ryan tells me about his surgery and subsequent recovery with the benefit and distance of hindsight. He is sanguine about the grief for his fertility that still hits at unexpected times—he describes being blindsided by emotion while playing with his new puppy—but above all, he is grateful that his hysterectomy brought relief from the pain and trauma of living in a body that was, in one hidden but crucial way, still female.

"Even after I'd been on testosterone for six months and my period had stopped," he recalls, "I'd still have nightmares that I was bleeding. I would wake up and my heart would just be racing. And after having the hysterectomy, those nightmares went away. It was just more and more freeing for me, with less anxiety weighing me down."

This is not to say that hysterectomy is always a positive experience, or that the journey to it is smooth. For many trans men, nothing could be further from the truth. In her book *Understanding Trans Health*, British sociologist Ruth Pearce writes a searing indictment of the gender identity services charged with enforcing the WPATH guidelines. She points to the fact that an individual must be seen (often by two or more professionals) to have the "disorder" of gender dysphoria diagnosed before being allowed to proceed toward gender confirmation therapies and treatments, and she argues, "The creation of gender identity disorders has worked to construct a professional class of gender identity *experts* [italics Pearce's], who may act as gatekeepers for trans-specific healthcare."[21] This sentiment was echoed by Dr. John Chisholm CBE, chair of the Medical Ethics Committee at the British Medical Association, when he testified at a 2021 inquiry into reform of the UK's Gender Recognition Act. "It's very onerous and dehumanizing to have to be asked all these intrusive questions in order to prove, in essence, that you are who you say you are," Chisholm argued. "We've come a long way from regarding gender dysphoria as a medical problem or a psychological problem or a mental health problem, and yet we are forced back into this paradigm by way of how the law operates."[22]

As I read Chisholm's words, I wonder what a gender identity specialist would think about me: would I be feminine enough, in my usual uniform of jeans and a hoodie, to qualify for state-sanctioned womanhood? Like so many people, I often rail against the gender expectations imposed upon me by society, and I don't always look, feel, or behave the way some people might think a woman should. I have a fractious relationship with my uterus—in some seasons of my life, pleading with it to give me children, and at others, cursing its propensity for heavy bleeding and pain—but what would a team of "experts" make of this relationship that veers from euphoric to dysphoric depending on the day, week, month, or year?

Ryan Sallans was able to work this often problematic system until, on the whole, it began to work for him. Ultimately, he tells me, he is grateful for every part of his journey, from Kimberly Ann and her fear of the dreaded "red note" to the tear-soaked puppy cuddles that still catch him unawares. He now makes a living from writing and speaking about the importance of choice and inclusivity in healthcare and the workplace, and he knows that he, like all of us, remains a work in progress.

"All this has allowed me to be able to give stories, and to teach people about things. Right? That's what my hope is," he says, "because I'm OK. I'm good."

-

As Ryan smiles and wishes me well, he gives the overwhelming impression of a man who is finally at peace with himself—a peace that is to be savored and shared and is all the more precious for the personal and physical battles that preceded it. He's happy and whole because he's lost a part of him that didn't fit. His journey demonstrates the intimate relationship between the womb and autonomy; the removal of his uterus enabled him to be his true self. He may be several centuries and thousands of miles away from poor Faith Raworth and her prolapsed womb, but for him, hysterectomy was just as desperately desired, and freedom from his uterus is freedom indeed.

Reprocide

RIGHTS AND INJUSTICE

*Had she no say in how the world made use of her? No way of
talking back to power? How much rage could the body house
before it exploded?*

—Saidiya Hartman, *Wayward Lives,*
Beautiful Experiments

No two hysterectomy stories are alike: each operation has its own reasons, whether to beat cancer, ease symptoms of disease, or affirm a sense of self. Thankfully, though, most of these tales have one abiding principle at their heart: that of informed choice. The decision to relinquish one's womb is, for the most part, freely made after counseling and consideration. Honor-bound to follow the Hippocratic oath to "do no harm" and by (one hopes) the pull of their own moral compass, most doctors raise their scalpels only when the patient's signature on the consent form is clear and dry.

Sadly, this autonomy has been denied to too many women around the world, now and in years past. History has shown that when a person's dignity and humanity are taken away, their reproductive rights are quick to follow. In these instances, the choice of what to do with one's uterus is usurped and perverted, and hysterectomy becomes a tool of mass oppression. This is especially true of women who have been marginalized by the dominant culture of the land in which they live: Black and Brown women; women whose religion is minoritized or maligned; enslaved, incarcerated, and detained women; and women who have been deemed "other," "less than," and undesirable for whatever baseless reason. Where systemic oppression resides, forced hysterectomy thrives.

The use of sexual violence as a weapon of war is as old as conflict itself. Rape features frequently in historical accounts of conflict, and it is alleged to have been used in or alongside battle by Persians, Israelites, Vikings, and Mongols, among many others. In *Our Bodies, Their Battlefield*, a searing account of this practice, Christina Lamb writes, "The word rape comes from the Middle English 'rapen,' 'rappen'—to abduct, steal, ravish, snatch. It originates from the Latin 'rapere' which means to steal, seize or carry away, as if women were property, which is exactly what men thought for so many centuries."[1] The advent of modern gynecology, with its own twisted roots in racism and misogyny, gave oppressors new ways of stealing women's power—not necessarily by rape in the traditional sense of penetrative sex, but by a different and equally insidious kind of bodily violation. The female reproductive system represented a new frontier to be conquered, and for those men with medical and social dominance, the uterus and surrounding structures became "their battlefield."

One of the foremost early innovators in the field, J. Marion Sims, refined some of the instruments and techniques still used today by operating on women already dehumanized by the monstrous system of slavery. Originally more of a general practitioner, Sims entered gynecology in the 1840s almost by accident, and with an undisguised distaste for women's bodies. "If there was anything I hated," he later wrote of his practice in Alabama, "it was investigating the organs of the female pelvis."[2] However, Sims's makeshift invention of a new kind of speculum for intimate examinations altered the course of his career forever. Called to assess a woman who appeared to have damaged her uterus after falling from a horse, Sims used a bent pewter spoon to open the patient's vagina and obtain a clearer view of her cervix, later boasting, "I saw everything as no man had ever seen before."[3] Suddenly, the female body was more than just a source of disgust, of little interest to the serious physician; with the introduction of that fateful spoon, Sims saw a new frontier on which he could stake his claim.

When neighbors brought Sims their enslaved girls and women, made incontinent by injuries they had sustained in childbirth and therefore "useless" in the fields, he embraced this new opportunity for gynecological exploration and experimentation with gusto. Numerous operations were performed on each of the young patient/victims, always without anesthesia, and often in front of a small crowd of other admiring physicians. The women's most intimate organs and most excruciating pain were on show for all to see. "Lucy's agony was extreme,"[4] Sims wrote of one of these tragic cases, and yet he persisted, trialing various methods of repair until he was satisfied that no further efforts could reasonably be made. Sims may have chronicled his career in detail, but the thoughts and wishes of Lucy, Anarcha, Betsey, and the other women whose bodies were fixed and flayed for all to see were never recorded. Their voices were silenced while Sims achieved lasting fame in his field: often feted as the "father" of modern gynecology, tools and techniques bearing his name remain in common use today. While some historians argue that Sims was simply doing his best within the unique social context of the time, his work was a dangerous conflation of medicine and oppression, creating a precedent for the assault of dehumanized and demeaned women under the socially acceptable guise of gynecological experimentation. This was a template that would be used time and again over the century to come.

Although slavery was abolished just as Sims's career drew to a close, another movement—as ugly in intention and as sweeping in ambition—began its inexorable rise. First coined by English intellectual Francis Galton, the term eugenics—from the Greek for "good origin," or "good birth"—described a movement that advocated the selective breeding of humans for the betterment of the species and the improvement of the world as a whole. In his 1869 book *Hereditary Genius*, Galton wrote, "As it is easy . . . to obtain by careful selection a permanent breed of dogs or horses gifted with peculiar powers of running, or of doing anything else, so it would be quite practicable to produce a highly gifted race of men by judicious marriages during several consecutive generations."[5] Theorists on both sides of the Atlantic embraced this idea, but eugenics met particularly fertile ground in America, where the double rise in emancipation and immigration posed a terrifying perceived threat to the security and identity of white middle- and upper-class society. Lady Liberty may have lifted her torch to welcome the huddled masses, but with the establishment of the Eugenics Record Office on Long Island in 1910, this insidious philosophy planted its flag firmly on American shores. The dark side of gynecology had been given the respectable veneer of a "movement," and reproductive control of marginalized women became enshrined in American society.

Over the coming decades, the reproductive rights of thousands of Black, Brown, poor, disabled, migrant, and generally "undesirable" women were decimated by a flurry of laws promoting eugenics and its evil handmaid, forced sterilization. By 1909, five states had passed or attempted to pass laws permitting the forced sterilization of people with mental disabilities, and in 1927, the Supreme Court case of *Buck v. Bell* opened the floodgates for this practice across the nation. At its center was Carrie Buck, a young woman from Charlottesville, Virginia, whose life was burdened by stigma and oppression long before her case captured the nation's imagination. Carrie's mother, Emma Buck, allegedly a prostitute with a history of drug abuse, rheumatism, pneumonia, and syphilis,[6] was deemed by the state to be intellectually inferior and subsequently committed to the Virginia State Colony for Epileptics and Feebleminded during her children's early years. Carrie was removed from Emma's care and soon taken in by local couple John and Alice Dobbs, but the adoption was ill-fated: in the summer of 1923, at the age of seventeen, Carrie was raped by Alice Dobbs's nephew, Clarence Garland.[7] Keen to dissociate themselves from the incident—which was blamed squarely on Carrie's "incorrigible behavior"—Mr. and Mrs. Dobbs sent Carrie the following January to the very same institution where her mother, Emma, still resided. Carrie's daughter, Vivian, was born in the Colony two months later, and in 1927, Carrie was used as a test case for legislators who argued that state-sanctioned sterilization was the only way to prevent such "feebleminded" women from perpetuating their flawed family line. Rather than looking on this beleaguered young woman with compassion, or even kindly pity, Justice Oliver Wendell Holmes Jr. ruled that Carrie—described as a "low-grade moron" typical of a "worthless class"—should be forcibly sterilized. "It is better for all the world," Holmes pronounced, "if instead of waiting to execute degenerate offspring for crime, or to let them starve for their imbecility, society can prevent those who are manifestly unfit from continuing their kind."[8] Carrie endured compulsory salpingectomy (removal of a uterine tube) in October 1927:[9] yet another violation of her dignity and autonomy, just five short months after Holmes announced that "three generations of imbeciles are enough."[10] Like so many cases before and after that brutal judgment, Carrie Buck bore the brunt of society's distaste while the man at the root of her misfortune escaped punishment; she later told biographer Paul Lombardo that Clarence Garland had promised to marry her after the assault, but that he then left Charlottesville, choosing instead to abandon his victim along with any responsibility for his crime.[11]

With the Supreme Court's fateful judgment, eugenics was enshrined in American law, legitimizing the legal but entirely nonconsensual sterilization of many thousands of women. Some, like Carrie Buck, were sterilized for being "feeble-minded," others were deemed dangerously promiscuous or abnormally attracted to partners of another race, and at one state-run institution in California, sterilizations were performed on women simply because they were deemed to have an abnormally large clitoris or labia.[12] Sometimes performed by hysterectomy, sometimes by cutting (often erroneously called "tying") the uterine tubes, the exact method is, to a large part, moot; both are an assault on the female reproductive system, and both procedures cut to the heart of a woman's power and identity. Many "reasons" were found, many techniques were used, and by the 1930s, America's enthusiastic and efficient adoption of forced sterilization had attracted the attention of history's most infamous proponent of eugenics: Adolf Hitler.

Long before his final rise to power as the leader of Germany's Third Reich, Hitler had admired America's commitment to perpetuating a superior race of humans. "There is today one state," he wrote in his 1924 manifesto, *Mein Kampf*, "in which at least weak beginnings toward a better conception (of immigration) are noticeable. Of course, it is not our model German republic, but the United States."[13] His remarks to a fellow Nazi demonstrate a particular interest in the ways in which American legislators had legitimized and systematized these dark impulses: "I have studied with great interest the laws of several American states concerning prevention of reproduction by people whose progeny would, in all probability, be of no value or be injurious to the racial stock."[14]

An estimated six million Jews and eleven million other persecuted people—ethnic Roma and Sinti, people with disabilities, gender nonconformists, political dissidents, union organizers, and so many others deemed undesirable—met their death at the hands of the Nazis and their collaborators. As in so many wars, rape, sexual abuse, and torture were commonplace; however, inspired by the systematic eugenics programs of America, Hitler's campaign of reproductive assault on persecuted women achieved an unprecedented scale of size and efficiency. Internment of Jews and others in concentration camps enabled death on an industrial scale; what is less widely known, perhaps, is the Nazis' focus on the wombs of incarcerated women.

True, history has often recorded the monstrous work of Josef Mengele, the doctor who carried out painful and humiliating sexual "experiments"—often including hysterectomy—on approximately fifteen hundred sets of twins at the Auschwitz concentration camp. Perhaps less well known, though, is Carl Clauberg, the gynecologist who was commissioned to design and implement a program of mass sterilization at Auschwitz and the all-female camp Ravensbrück. In a 1943 letter to Heinrich Himmler, one of the senior architects of the Holocaust, Clauberg enthusiastically described a new method that could "achieve the sterilization of the female organism without operation."[15] The terminology itself reveals the dehumanization that made the Nazis' abuses possible—the use of the impersonal, animalistic "organism," for example, rather than "person" or "woman"—and the desire to curb the reproduction of undesirable populations as cheaply and easily as possible ("without operation"). Clauberg goes on to describe his novel method, a gruesome combination of uterine injections and X-ray radiation of the pelvis, and he closes his letter by boasting that he will easily meet the projected targets given to him earlier by Himmler: "One adequately trained physician in one adequately equipped place . . . will most likely be able to deal with several hundred, if not even 1,000 per day."[16]

One of Clauberg's survivors later confirmed the use of these methods with her heartbreaking testimony, saying, "I cannot recall one woman who had agreed to any such experiments—to the contrary. Dr. Clauberg performed sterilization experiments on my person without my consent." In spite of the excruciating pain and indignity of repeated uterine "treatments" over many months—grueling episodes in which she was restrained and medically tortured—the woman recalls, "I did not protest because it would have been senseless. It happened anyway."[17]

I appreciate that it is difficult to read even a vague summary of these events, and it is tempting to skim past them. As someone whose extended family was decimated by Nazis during the Holocaust, I, too, would rather not consider these atrocities in detail. However, it is essential to understand the way in which the uterus was uniquely targeted by Hitler's regime, not only so that we may recognize and honor the pain experienced by the Nazis' victims, but also so that we can recognize the myriad ways in which this legacy of forced sterilization has continued to evolve and exist in the so-called civilized world over the intervening decades.

First impressions appear to show America backtracking from its eugenic stance. In 1942, the Supreme Court—cognizant of the atrocities happening overseas—began to lose its nerve around forced sterilization, ruling in *Skinner v. Oklahoma* that sterilization could not be used as a punitive measure against certain convicted criminals.[18] However, there is evidence to suggest that coercive sterilization was still being used against Japanese American women interned in American-run camps during the Second World War,[19] and even after the war had ended, the impulse to curtail and control marginalized women's reproductive lives simply evolved with the times. In the postwar years, programs of coercive sterilization pushed back against the increasing diversification of the American population. Hospitals, legislators, and insurance companies found new ways to target women who stood to gain equality and dignity; sterilization was sometimes legitimized (as in a campaign that saw 30 percent of the women in Puerto Rico sterilized by 1965, and over three thousand Native American women sterilized without consent in the 1970s), often incentivized (with Medicaid paying doctors far more to perform hysterectomy than tubal ligation), and widely used as a training opportunity for inner-city medics.[20]

In the early 1960s, performing hysterectomies on Black women attending hospital for ostensibly minor procedures was so widespread that this kind of forced sterilization became colloquially known as a "Mississippi appendectomy,"[21] while in the North, the head of obstetrics and gynecology at one hospital admitted, "In most major teaching hospitals in New York City, it is the unwritten policy to do elective hysterectomies on poor Black and Puerto Rican women, with minimal indications, to train residents."[22] Similar reports emerged from Boston and Los Angeles. These operating theaters may have been miles away from the Nazi camps, but morally, there is little to separate Carl Clauberg from the ease with which the American medical establishment devalued and destroyed the wombs of marginalized women. Hysterectomy continued to be used as a tool of population control and a weapon of white, patriarchal supremacy. As the civil rights movement marched on through the 1960s and the feminist movement gained pace in the 1970s and '80s, certain women were still being treated as "organisms" whose reproductive capacity represented a clear and present danger to those in power. Even in the new millennium—with the first Black president and vice president, and a tide turning toward the "liberty and justice for all" promised by the national pledge of allegiance—this grim chapter of American history is still unfolding.

> *African slavery, as it exists in the United States, is a moral, a social, and a political blessing . . . You cannot transform the negro into anything one-tenth as useful or as good as what slavery enables them to be.*[23,24]

These are the words of Jefferson Davis, the American president who led the South throughout the Civil War and whose memory is honored at the sprawling Jefferson Davis Memorial Historical Site about 150 miles southwest of Atlanta, Georgia. Visitors to the thirteen-acre site can enjoy a museum with carefully curated cases of Confederate weaponry, flags, and uniforms; a nature trail through pleasant woodland; a gift shop; and a statue of the man himself, erected by the United Daughters of the Confederacy on the exact spot where the leader of the Southern states was captured by Union troops on May 10, 1865. A video on the park's website describes this raid and the subsequent collapse of the Confederacy in slow, baleful tones as nothing less than "the end of a dream."[25]

The Confederacy and its leader's dreams may have melted in the dawn mist of that morning in 1865, but a short drive away from that quiet spot in the woods, proof of the ongoing persecution and dehumanization of Black and Brown people survives in a complex of long, low concrete buildings surrounded by a barbed-wire fence too high for hands or dreams to climb. At 132 Cotton Drive—its very address bearing the vestiges of America's problematic past—Irwin Detention Center is one of the many sites built to accommodate the human fallout of the Illegal Immigration Reform and Immigrant Responsibility Act of 1996, which delegated the identification and detention of "aliens" (non-native migrants) to state and local agencies.[26] Like most of these centers, Irwin was established and run by a private, for-profit company. Detention is big business: Irwin alone could accommodate over 1,200 people detained by ICE (the Immigration and

Customs Enforcement agency), a small fraction of the over 200,000 foreign-born men, women, and children migrating to America each year. Like other centers, Irwin has housed a veritable United Nations of refugees and asylum seekers, hailing from Mexico, Nigeria, Guatemala, Nepal, Cameroon, Pakistan, China, and beyond. Like other such facilities, Irwin's bleak, bland exterior belies the diversity of human experience that exists within its concrete walls. Unlike other facilities, Irwin has achieved notoriety for the rampant reproductive violence alleged to have been inflicted on its female residents. According to a major class-action lawsuit incorporating the testimony of over forty women, Irwin detainees were routinely sent to and forcibly sterilized by a man known to his victims as "the uterus collector."[27]

Azadeh Shahshahani knows this gruesome story better than almost anyone else. Formerly the director of the National Security/Immigrants' Rights Project for the American Civil Liberties Union, Azadeh is now the legal and advocacy director for Project South, a Georgia-based activist organization. Azadeh is also an immigrant herself, having fled with her family from Iran to the US at the age of fifteen. When we speak during a late-night video call, our conversation—one migrant to another—feels intimate and immediate in spite of the ocean between us.

"It was about mid-2010 or so when I started going to Irwin," Azadeh recalls, "because they started detaining immigrants there about a decade ago. At the beginning, they were allowing me to have contact visits with the detained people. And then at some point, they found out that I'm a human rights lawyer," she says with a rueful smile, "so then those visits started happening through a piece of Plexiglas." The authorities had good reason for concern: Azadeh's visits often focused on the detainees' numerous allegations of mistreatment at Irwin, including inedible food, limited access to visits and calls from family and lawyers, and hard labor with minimal pay.

Poor healthcare became a running theme in these conversations. Female detainees disclosed inadequate gynecological and antenatal care, and a 2017 report from Project South documented what appeared to be a blatant disregard for intimate health.[28]

"One issue that posed particular concern," Azadeh tells me, "was lack of access to clean underwear for women. They were being given used underwear and wet underwear for a long time. And I brought this to the attention of the warden . . . and the fact that this is a huge concern, and she basically viewed it as a nonissue. You know, no problem at all. And I think the only way you could do that is if you engage in dehumanization of the detained immigrant population."

The shocking extent of this dehumanization was revealed in 2020 when Dawn Wooten, a nurse at Irwin Detention Center, turned whistle-blower. A report from Project South incorporating testimony from Wooten and detained individuals chronicled a dizzying range of abuses at Irwin, including falsification of medical records, inadequate Covid control measures, and generally unsanitary conditions. Most alarmingly, Wooten reported her concerns about the disproportionately high rate of hysterectomies performed on Irwin detainees by a local doctor, Mahendra Amin. According to Wooten's complaint, many women had not consented to the procedure, often only discovering after the fact that it had even been performed during what should have been exploratory or minor surgery.

While accepting that some women may have been experiencing heavy bleeding or other conditions necessitating hysterectomy, Wooten suggested that "everyone's uterus cannot be that bad" and described the increasing alarm among her nursing colleagues about Dr. Amin's practices. "We've questioned among ourselves, like goodness, he's taking everybody's stuff out . . . That's his specialty, he's the uterus collector."[29] One detainee said that she was aware of five different women who had had hysterectomies at Amin's clinic between October and December 2019 alone, saying, "When I met all these women who had had surgeries, I thought this was like an experimental concentration camp. It was like they're experimenting with our bodies."[30]

With the weary disappointment of someone who has witnessed a catalog of horrors throughout her career, Azadeh tells me that she was shocked but not surprised by these allegations of sterilization by stealth.

"I think it was a manifestation of the dehumanization we had already witnessed," she says, "but given the broader context of everything that we had seen at Irwin over the years, and the level of impunity . . . it was not surprising."

As the allegations emerged and Project South scrambled to gather testimony from women who had witnessed or experienced firsthand this suspected reproductive abuse, Azadeh met with another unfortunate but unsurprising obstacle: resistance and obstruction from the immigration agency itself.

"ICE obviously did not cooperate in any way in the investigation," she says. "If anything, they were trying to cover their own tracks by deporting witnesses and survivors. One of the women who played a really fundamental role in providing her testimony as part of the complaint was 'Hiromi.' After the complaint started getting the attention that it did, [ICE] basically asked immigrants, you know, 'Who spoke to the lawyers?' and she came forward and said, 'I was one of the people who did.' And the next thing she knew, she was on a plane—they deported her right away . . . The retaliation was swift. They deported six people before Congress and lawyers stepped in to stop the additional deportation of witnesses and survivors."

Undeterred by this resistance, Azadeh and her colleagues followed word-of-mouth leads to track down dozens more women who had left or been deported from Irwin, ultimately using this testimony in a class-action lawsuit against ICE, LaSalle Corrections, Mahendra Amin, and others. The backlash was almost as swift and vigorous as ICE's first retaliations: many residents of the Ocilla area argued that Amin was a respected member of the local community who had cared for many of its women and delivered many of their babies with competence and compassion. A "We Stand with Mahendra Amin" Facebook group garnered over fifteen hundred followers, with stories from Amin's long career featured alongside photos of supporters bearing #TeamAmin signs and wearing pro-Amin T-shirts.[31] The Irwin County Hospital issued its own statement of unequivocal support, arguing that Amin was "a longtime member of the Irwin County Hospital medical staff and has been in good standing for the entirety of his service to the Irwin County community."[32]

At the time of writing, the class action has not yet come to trial. Amin is still practicing medicine and has not faced any professional recriminations. Azadeh tells me, "The only thing that ICE was forced to do was to stop sending the detained immigrant women to him. And they definitely dragged their feet. That didn't happen immediately after our complaint came out." She is quick to emphasize, though, that the problem of forced sterilization of ICE detainees is likely to exceed the scope of a single doctor's practice. Rather, it could well be happening on a much wider scale.

"This is a systematic problem," she says. "This is not just one individual." To that end, Project South has filed Freedom of Information requests to investigate the potential extent of the issue in other states and institutions. Azadeh is understandably circumspect in her discussion of these investigations; in typically oblique lawyer-ese, she tells me, "We have received some information. We are currently analyzing it." She does concede, though, that "I would not be surprised if similar atrocities are happening at other detention centers around the country, given what we know about what's been happening, especially at corporate-run detention centers."

Other similarly chilling stories from America's recent past suggest that state-sanctioned hysterectomies may be far from isolated incidents. In California, an investigation prompted by whistle-blowing from incarcerated women revealed that as many as fourteen hundred forced or coercive sterilizations may have been performed on inmates between 1997 and 2013.[33] As with the allegations in the Project South complaint, women were sterilized during what they thought were investigative procedures or treatments for minor gynecological issues.

"We actually used to call them surgeries of the month," says one inmate in *Belly of the Beast*, a documentary chronicling the California investigation. "That was the cure-all. That's what it was."

The film's director, Erika Cohn, tells me, "It all comes down to whose lives—whose wombs—are of value," and says that although the state has now recognized these abuses, it has a long way to go toward closing legal loopholes and achieving justice for the women whose reproductive rights were so cruelly denied. "We need accountability for these eugenic practices, justice for the survivors, and safeguards to prevent future abuses," she says, alluding to a reparations bill that's making its slow, steady way through California's legislature at the time of writing. "Eugenics is alive and well. We see systemic racism and population control through so many facets: policing, imprisonment, the immigration detention system, who has access to healthcare and education, etc."[34]

Reports of reproductive abuse continue to emerge from every corner of the globe, from the state-sanctioned and incentivized sterilization of Roma women in Czechoslovakia (and later the Czech Republic) from 1966 to 2012,[35] to widespread ongoing allegations of forced hysterectomy and coercive contraception among Muslim Uyghur women in China.[36] Author and reproductive justice activist Loretta Ross has a grimly apt name for this phenomenon. "I coined the term 'reprocide' to describe when genocide is primarily committed through reproductive control," she writes.[37] According to Ross—and as our journey from Sims to Hitler to the "uterus collector" and beyond attests—reprocide never dies, it just evolves across time and space according to the needs of the dominant culture. "Black women know population control ideologies morph over time," she says, "but are never totally abandoned."[38]

Reprocide, then, is not new. The uterus may as well be marked with a bloody bull's-eye, so powerful and precious a target it is for any regime wishing to oppress the people whose very existence poses a threat to the status quo. More often than not, these people are biologically female; men have certainly been victims of forced sterilization, too—true barbarism is undiscerning—but they have not suffered reprocide in as many ways, in so many places, and over so many years as women have. One could easily be forgiven for adopting a fatalistic view in light of these ongoing abuses; however, even Azadeh Shahshahani remains cautiously optimistic that righteousness will prevail.

"This work is exhausting," she admits as our conversation draws to a close. Her voice is still inflected with the elegant vowels of her Iranian native tongue, every word a reminder of her roots as well as her convictions. "You've just got to keep at it, though," she tells me. "Because, you know, as Martin Luther King said, the arc of history is long, but it bends toward justice.[39] And that is absolutely true."

The week after our call, Azadeh shares exciting news on social media:

> *BREAKING! Immigrants will no longer be detained at the*
> *corporate-run Irwin County Detention Center, where women*
> *were subjected to medical abuse. This momentous victory is the*
> *result of years of organizing and exposing the human rights*
> *violations by orgs on the ground. Onward!*[40]

It takes me a moment to absorb the enormity of this statement. I can't help but feel that—against all odds—the arc Azadeh described during our conversation has just curved a degree or two closer to justice. That curve can be so easily deflected, and that justice is so fragile: it depends on a society that honors bodily autonomy and respects the womb as the nexus of reproductive freedom, regardless of whether that womb resides in a body that's Black or Brown, rich or poor, incarcerated or free. There is still work to be done, with challenges yet unseen. With every generation, the trajectory of new life arcs into a future as yet unimaginable.

Future

INNOVATION AND AUTONOMY

O brave new world,
That has such people in't!

—WILLIAM SHAKESPEARE, *THE TEMPEST*

It's six o'clock on a pitch-black November morning, and I'm stand-ing on the steps of the women's health clinic at the University of Gothenburg in Sweden. The lights of the *Kvinnokliniken*'s foyer spill a cool fluorescent pool onto the darkness of the steps outside, and I stop there for a moment to remind myself why I've come: to catch a glimpse of the future of the uterus, and to meet the man who's shap-ing it. Mats Brännström is professor and chief physician of obstetrics and gynecology at the university's Sahlgrenska Institute, and he led the team that performed the world's first successful uterus transplant; that is to say, one which resulted in the subsequent birth of a live baby.[1] Since that groundbreaking procedure, the Gothenburg team have repeated the procedure several times, even introducing the use of remote-controlled robotic instruments for less invasive surgery.[2] Womb transplants—the kind of futuristic reproductive swaps that most people would think were impossible—are Dr. Brännström's specialty, and I'm here to watch the master at work.

As I enter the clinic building and climb the stairs toward the Reproduktionsmed department, I'm troubled by the niggling feeling that I'm here to meet the wrong person. I've been learning as much as I can about Dr. Brännström—his background, his achievements, his plans for the future—and I've watched so many of his lectures and interviews on YouTube that I feel like I already know him. Someone else wrote the beginning of this story, though: a woman on the other side of the world. I know her name, but not her face. It's "Angela," or at least, that's what Dr. Brännström calls her. I should, I think, be meeting Angela.

I know very little about her, but what I do know is that her offhand comment set in motion a chain of events that includes two decades of womb-swapping, an international race to refine and surpass techniques that seem like the stuff of sci-fi movies and today's journey to the Sahlgrenska Hospital. In 1998, Mats Brännström embarked on a long journey, too—to Adelaide, Australia, where he planned to advance his work on ovarian function by specializing in the study and treatment of infertility. As luck would have it, the only research space available was in gynae-oncology—the study of cancers of the female reproductive system.

And that's where Angela comes in. A cervical cancer patient, Angela's treatment included a hysterectomy—a procedure that would help prevent any spread or recurrence of her cancer, but would also render her unable to gestate and birth her own child. Although adoption was an option for women in her position and even surrogacy was in its early days of popularity, Angela wasn't satisfied with these choices. She pointed out to Dr. Brännström that organ transplant had become a widely accepted part of modern medicine, with increasing rates of success—so why couldn't he transplant a uterus?[3,4]

"I thought she was a bit crazy," Dr. Brännström has said,[5] so the young medic did what many men do when a woman makes a comment so left-field, so irrational, so nutty that it illuminates a million wildly flashing lightbulbs in your mind—he went to the pub and told his mates. The more he discussed Angela's suggestion with his colleagues, the more he realized that a uterus transplant was "a very good idea."[6] That's where Angela leaves the picture, though; Dr. Brännström soon realized that before he could start playing a game of musical wombs, he would need to prove the procedure's safety and efficacy.

As it happens, there had already been an attempt at human uterus transplant before Brännström was even born. In 1931, Danish transgender artist Lili Elbe received a uterus during an operation at the Women's Clinic in Dresden. While Elbe had already undergone several risky surgeries—including removal of her penis, and implantation of ovarian tissue—to achieve her goal of leading a fully female life, this final, pioneering operation was a step too far. Her heart may have been set on motherhood, but without the benefit of today's sophisticated immunosuppressant drugs, Elbe's body could not fulfill that dream. Infection set in, and three months after the transplant, Lili Elbe died of a cardiac arrest.[7]

After such a tragic outcome—and with trans health occupying the tiniest niche in conventional medicine—the scientific community appears to have abandoned human uterus transplants over subsequent decades. Every now and then, researchers risked a brief flirtation with the idea; in the 1960s, scientists at the University of Mississippi trialed a canine womb transplant, and the recipient even managed to sustain a successful pregnancy.[8] But like so many medical advances that hold promise for human females, further development of the procedure appears to have been scant after that initial success. Perhaps this kind of surgery was regarded as a curiosity, or an aberration of nature; certainly, there doesn't seem to be any recorded discussion of the significance of uterine transplants for the women of the world, 1.5 million of whom are estimated to suffer from absolute uterine factor infertility (AUFI)—essentially, infertility caused by the lack or dysfunction of one's own uterus.[9]

Fast-forward to Angela and her "crazy" suggestion, and then quickly onward to young Dr. Brännström, who began to try womb transplants in rodents, sheep, pigs, and—during a formative trip to Kenya, where the rules on animal experimentation were more lax at the time than those in Sweden—baboons. And from these African primates, Brännström made the inevitable jump to their human cousins—women whose names and faces are hidden from the public domain, but whose willingness to either donate or receive a uterus has earned them a place in medical history.

For the rest of the day, as the darkness of a Swedish morning melts into milky daylight and back into dusky gloom, I watch history unfold on a large screen in a conference room next to Mats Brännström's operating theater. Joining me around the table are medics from across the globe, many of whom have recently attended a conference in Cleveland where Dr. Brännström shared his findings. An elite group, it seems, have been invited here to observe this robot-assisted transplant so that they can bring the knowledge and the technique back to their own countries, where clinical trials are in various stages of completion. The Taiwanese surgeon across the table tells me that uterus transplants could be an important option in countries like his where surrogacy is illegal; the American obstetrician at the top of the table looks unimpressed, having established her own transplant program already; and a friendly Australian doctor to my right asks me if I'm staying until Monday to "do the sheep"—apparently, the doctors have been invited on a kind of package holiday, starting with today's operation and ending with a hands-on workshop in which participants will swap animal wombs under the tutelage of Dr. Brännström himself.

I may not have been invited to "do the sheep," but I do share the doctors' close-up view of some futuristic uterine magic. Leaflets on the table tell us about the women involved in today's transplant. The donor, I read, is a thirty-seven-year-old mother of three; most important, her children have all been born vaginally, so her uterus is still intact and undamaged by surgical scarring. The recipient, age twenty-one, is her sister. Apparently, hers is one of the most common reasons for absolute uterine factor infertility: she has Mayer-Rokitansky-Küster-Hauser syndrome (MRKH), a congenital condition in which the vagina and uterus are underdeveloped or absent.[10] Many women with MRKH (or #MRKHWarriors, as the small but vocal online community often calls itself) only realize their bodies are different in their teens when, unlike their peers, their period never arrives. Sexual relationships can pose further problems; this seems to be when many women with MRKH pursue medical assistance, with varying degrees of success based on the knowledge and bedside manner of the medic in question. Blogs and forums are bursting with stories of doctors openly googling the condition, doctors making prurient and distressing comments about women's intimate lives, doctors blundering through treatment plans involving the use of progressively larger vaginal dilators—the potential for humiliation bordering on abuse seems boundless. While some of these self-proclaimed "warriors" become mothers through adoption or surrogacy, what's clear is that all of them have fought a lifelong battle against ignorance, pain, and misunderstanding. It's little wonder that many women with MRKH are enthusiastic about the hope offered by uterus transplant trials, in spite of the rigorous screening and potential risks; what lies ahead could hardly be any more of a challenge than what's already been endured.

For those of us gathered today in Gothenburg, the transplant represents modern medicine's best efforts to overcome the human body's challenges. For nearly ten hours, we watch Brännström and his team probe the donor's pelvis, each move magnified by a laparoscopic camera inside the body. The work is painstaking: for the uterus to be removed successfully, ligaments and ureters must be moved with minimal disturbance. The tiniest vessels of the uterus must be divided one by one, and arteries must be clipped with expert precision.

"It's not like a usual hysterectomy, where you can just cut everything off and pull the uterus out through the vagina and toss it in the bin," the friendly Aussie on my right whispers when I marvel at the detail of the procedure. "If you're going to use the uterus again, you need to make sure everything is intact and functional. It's so much more challenging." That much is becoming clear.

Finally, there is a moment in the early evening when the mood of the group sharpens in quiet anticipation. Inside our strange little cinema there is a clarity of air, a refocusing of eyes; I recognize it as the same subtle shift that occurs in the birth room when, after many hours of seemingly fruitless pain during which even the mother has almost forgotten why she's there, a powerful contraction nudges forward the first visible sliver of a baby's head.

On the big screens, the donor's uterus has been cut almost completely free of its venous entanglements, and the fleshy cervix joining the womb to the vagina is about to be detached. We sit forward in our seats, wide-eyed; our pupils dilate as one. Between us, we have spent thousands of hours in operating theaters; we edge toward the screen until we can almost believe that we're inside it, scrubbed up and gloved. At 5:20 p.m., the air in the room changes again; it crackles with the electricity of birth. The uterus is out. The camera pans back from the donor's abdomen and we see the organ held aloft by a surgeon's disembodied, blood-streaked hand. I've seen something like this moment many times during Cesarean sections—the moment of delivery, when a goo-slicked, tumble-limbed child is lifted out, its umbilical cord still spiraling into the wet, hot hole of the open abdomen below—but in this case, there is no baby, and no cry. This delivery holds only the promise of a baby to come, and the echo of its cry.

The moment of triumph is short. Our screen goes blank; evidently there is no camera at the "back table," the area in the operating theater where the uterus will be inspected, cleaned, and flushed through with anticlotting drugs before it is deemed fit for transplantation into the recipient's body. In the conference room, we fidget and sigh and shuffle our papers. Some of us have tears welling in the corners of our eyes; some of us have been holding our breaths, and now we exhale and tut and feel exposed, like when the lights come on at the end of a weepy movie and no one wants to be caught crying.

As it happens, the actual transplantation of the womb into the donor is much quicker than the operation that preceded it. Our screens flicker on, and there is the recipient's body: identical to her sister's on the inside—ureters and ligaments and smooth, glistening bowel—but for the absence of a uterus. This is about to change, though, because here is the prized organ, blanched and lifeless. A vascular surgeon works with painstaking care to graft the uterus to its new blood supply; tiny nicks and joins and sutures are made in these vessels deep within the pelvis, all the more remarkable when I remember that what I'm seeing on the screen has been magnified many times by the camera's lens.

And then, a moment of blindsiding power: the first graft has been prepared and the connection is ready to be tested. A clamp is released from a newly connected artery. In the space of a heartbeat, fresh blood rushes into the adjoining vessel and a pink blush spreads slowly across the uterus—until now, a fist of pale white muscle. I can hardly believe what I'm seeing. What was dead now has life; what was still and cold now pulses with heat. In this moment, the echo of that future baby's cry seems to grow a little louder. The impossible seems—incredibly and against all odds—possible.

THE SIGHT OF A LIFELESS MUSCLE SURGING WITH VITAL-
ity is almost too strange to be believed, but the even stranger truth is
that in the future, fertility treatment for people without wombs may
not require any womb at all—or at least, not wombs as we currently
know them. Even as the field of uterus transplantation emerged and
clinical trials sprang up on almost every continent, scientists were already thinking one step ahead, envisioning a world in which synthetically produced wombs could eliminate the medical and psychological
complexities of human organ donation. In 2016, Mats Hellström, an
associate professor of bioengineering and organ regeneration at the
University of Gothenburg, led a team that successfully used scaffold-
like "patches" of lab-grown tissue to repair uterine damage in rats; the
goal, as Hellström later described it, was "to create a bioengineered
organ to replace the need for a donor."[11] The race to develop a functional, lab-grown uterus has continued apace since then. Four years
after the Swedes published their findings, Dr. Anthony Atala, the
director of the Wake Forest Institute for Regenerative Medicine in
North Carolina, used a slightly different technique to create repara-
tive patches of tissue for rabbit uteruses, resulting in the safe gestation
and birth of several baby bunnies:[12] a clinical but no less astounding
version of a magician drawing a rabbit from a hat.

Elsewhere, some scientists eschewed human tissue altogether
in their pursuit of uterine advancement; in 2017, researchers at the
Children's Hospital of Philadelphia revealed they had safely nurtured extremely preterm lamb fetuses—roughly equivalent in age to
twenty-three-week human fetuses on the very edge of viability—in
plastic "biobags."[13] News outlets around the world jumped on the
story, featuring eerie photos of a baby lamb floating in its biobag of
synthetic amniotic fluid, surrounded by a tangle of tubes and valves.

The ability to gestate a human fetus outside the womb from conception to full-term birth may yet be decades away; in the meantime, the image of a lamb sleeping peacefully in its biobag provides a vision of a potential future that is as compelling as it is unsettling.

This vision could not be more different from the kaleidoscope of humanity that spins through my days as a midwife. In that role, I am surrounded by the noise, heat, and color of women's bodies in all stages of procreation. Each hour that passes is saturated with the stuff of birth: the blood that trickles, rushes, pours, and clots; the amniotic fluid whose salty-sweet smell lingers on my skin long after my shift has finished. There are occasional tragedies when nature surprises me with sickness and death, but more often, there is triumph, joy, and the reaffirmation of the womb's power as the seat and source of new life. I wonder if midwives in this new age of biobags will be little more than lab technicians, walking down endless aisles of fluid-filled tanks, monitoring the hundreds of silent, sleeping babies suspended in their man-made, watery worlds. Birth as we know it—visceral, ugly-beautiful, and raw—might become little more than a distant memory in human consciousness, a quaint practice of primitive people. No more need for the uterus with its unpredictable cycles and inconvenient emissions, no more "obstetrical chairs" and labor wards, no more blood, no more mess. What will become of midwives if labor becomes a bygone chore? More to the point, what value will be placed on a birthing body that does not birth, or on the womb itself, an organ of flesh and flaws whose evolutionary purpose has been made redundant, replaced by the daring new synergy of man and machine?

Aldous Huxley provides a hypothetical answer to these questions in *Brave New World*, a novel in which society has "advanced" to the point at which sex and the burden of pregnancy have been completely separated. In Huxley's world, babies are gestated in bottles and decanted—rather than birthed—at the time of delivery. Women

261

who still endure pregnancy the old-fashioned way are described as "having children all the time—like dogs. It's too revolting."[14] To use the female body in its most essential reproductive capacity is, in this "perfect" society, primitive and distasteful. To choose pregnancy and birth, with all the attendant mess, bodily chaos, and indignity, is to debase oneself beyond redemption.

So far, so science fiction. Perhaps this strange world of bottled babies and clean, disembodied birth is actually a vision of progress: one in which biological parenthood is available to any person who desires it, and one which could, in turn, free those people from the "burden" of reproduction. Synthetic wombs may be the key to a society in which, at last, the sexes are truly equal; when breeding can be outsourced to the lab, any man or woman can work, play, and live life to its full potential without the time-consuming (and messy) interruption of pregnancy and birth. Writing about the advent of novel womb technologies, philosophers Elselijn Kingma and Suki Finn paint a tantalizing picture of a society in which this imbalance is eradicated: "[A]lthough the promise of a genuine artificial womb remains sci-fi, its lure is understandable; what pregnant person has not wished— albeit only briefly—that they could leave their 'body in its bulk and weight'; or that she could 'park her fetus on a shelf'—and run, drink, smoke, jump, dance, work or make love ad libitum, free from the risks, burdens, and moral and physical constraints that actual gestation entails?"[15]

That wish for emancipation may be nurtured, secretly or otherwise, by every pregnant person who has ever felt the pang of full-term heartburn, or concealed their growing bump from an unsympathetic employer, or simply longed for a break from years of near continuous childbearing. The possibility to pause some or all of one's pregnancy— fetus percolating quietly away in its biobag all the while—may be tempting, but it is not without significant bioethical dilemmas.

Chloe Romanis, assistant professor in biolaw at Durham University, argues that artificial wombs may represent a kind of reproductive Trojan horse: outwardly irresistible, but on closer inspection, fraught with danger.

"I think the male gaze would always favor gestation that's external, that can be measured, that can be controlled and viewed," Chloe tells me.

We are speaking via video link while her dachshund, Nora, clambers onto her lap and my own daughter works in the next room; as is customary for these interviews, we each apologize for our domestic distractions. Undeterred by Nora's enthusiastic interruption, Chloe explains that in a world dominated by the male gaze, artificial wombs—measurable, controllable—may be deemed vastly preferable to the flawed, unpredictable, and morally ambiguous bodies and behaviors of flesh-and-blood women.

"I think there's still a sense that women can't be trusted," she says, citing the patriarchal need to police pregnant women's actions, from the foods they consume to the work they do and the partners they have. While I was writing this book, many women contacted me to share their own stories of this insidious behavioral correction. Everything from caffeine to alcohol to clothes and exercise is subject to scrutiny by family, friends, and strangers; comments range from the lighthearted to the more insidiously aggressive.

"There's a sense that the formation of human beings is a magical process, and we need to make sure that these women are doing everything right," says Chloe. In the future, she suggests, a woman whose behavior does not conform to social norms may be threatened with forcible preterm delivery, with the fetus then being transferred to the "safer" artificial womb for the remainder of its gestation.

"I think the argument would go something like this: removing that fetus from that pregnant person increases the chance of that fetus having a healthy life when it's born. Therefore, we should do that. It's almost a subtle form of coercion," she explains, "where it's like, listen, there's this machine that could do it perfectly. Therefore, you have to live up to the standard."

Chloe calls this partially external gestation "ectogenesis"—a kind of Greek portmanteau meaning, quite literally, "creation outside"— and in a 2020 piece for the *Journal of Medical Ethics*, she and her coauthors warn that "pernicious narratives about control, conflict and the womb must be addressed in the face of these technological developments."[16]

While acknowledging that ectogenesis may be potentially life-saving for extremely preterm fetuses facing threatened miscarriage or other imminent compromise, Dr. Romanis argues that there is a large and ethically dangerous gray area in which the risks and benefits of human gestation are subjective. The potential for exploitation within this gestational twilight zone is enormous.

"Coercive intervention in birthing happens all the time," she says, "and it's incredibly likely that this technology just fuels that."

It may be uncomfortable to imagine that birthing people are subject to undue pressure and influence at such a formative and vulnerable time in their lives, but research suggests that such coercion is a pervasive element of the broader phenomenon of obstetric violence. A Swiss study published in 2021 found that, of 6,054 women, 26.7 percent—more than one in four—had experienced some form of informal coercion during labor and birth. Notably, the rate was even higher among migrant women and women who wished to have a vaginal delivery.[17] I am ashamed to admit that I have witnessed this behavior in my own practice; although most of the encounters I see in the course of my daily work are respectful and empowering, I have also seen many women swayed by practitioners and partners to make a choice—whether to take or avoid pain relief, to opt for Cesarean or vaginal birth, or to accept or decline a certain kind of monitoring or medication—on the basis that it is best or easiest or quickest, without full discussion of the risks, benefits, and alternatives to doing so. In this context, it is easy to imagine a near-future scenario in which women are coerced into considering ectogenesis for a variety of clinical or ethical reasons, spurious or otherwise. Who is to say what a mother would do if her doctor insisted that a biobag would be "best" for baby, or if a court threatened a woman with ectogenesis if she did not conform to certain standards of behavior during pregnancy?

While a world in which babies are forcibly delivered from the wombs of wayward mothers may seem reassuringly distant, Chloe argues that reproductive technologies are already being used to coerce women into using their wombs in certain ways. Ostensibly, these transactions are presented as "perks" for the women themselves, but on closer examination, they serve to benefit the women's employers and the wider capitalist system in which they exist.

"For example," she says, "Facebook, Apple, and Google offer female employees egg freezing. So they say, 'Look, one of the benefits of working for us is we'll freeze your eggs so you can be fertile for longer.' If this [artificial womb] technology exists, they would have real incentives to go, 'Listen, we respect that you want kids, but why don't you just not be pregnant?'" Chloe suggests that it's only a matter of time before ectogenesis becomes another executive perk. "I mean, it sounds melodramatic," she says, "but as long as we live in a capitalist society, there are reasons why people would be pressured into using these machines."

If and when artificial wombs become practical, effective, and affordable, will their use usher in a new two-tier society of gestators and non-gestators? In a world in which blue-chip companies can rent lab space for their female employees' fetuses, will a biobag baby be the next executive status symbol—a logical step up from the company car and the corner office? And away from the gleam of the boardroom's glass and chrome, in this brave new world's darker, less salubrious sectors, will the constant threat of the scalpel hang over women with socially undesirable behaviors? Will the early Cesarean section and subsequent transfer to a clean and controllable bio-womb be deemed medically and morally safer than continued gestation in the body of a dangerously flawed human? Or—if full-term ectogenesis becomes a feasible reality—will this technology offer precious hope to those couples who are unable to sustain a pregnancy themselves, and might otherwise have chosen a surrogate to carry and birth their child? Same-sex couples, single parents, women whose health precludes a pregnancy—will synthetic wombs provide a safe alternative "space" in which to nurture and grow embryos conceived by IVF?

At the time of writing, these questions are moot, and the ethical conundrums posed by artificial womb technology are likely to remain hypothetical for some years to come. Chloe admits that she's often asked why she is preoccupied with an alternate uterine reality that may not come to fruition within her lifetime, if ever.

"[People say] why do you write about this all the time? Why do you concentrate on this thing? And to me, it's because it's an imaginary [scenario] that can help us to think about how we value the womb now. I think those questions are just fascinating, even if they're entirely abstract."

Later that afternoon, my daily wander down the internet's womb-related warrens leads me to a photograph of something that, at first, looks distinctly industrial: inside a glass box strewn with cables and switches lies a multichambered wheel. With more than a passing resemblance to the barrel of a large gun, this wheel holds a vial within each of its gently spinning apertures, and within each vial is a mouse embryo floating in pseudo-amniotic soup. These mice, though, are not like other mice, and their mechanical trappings represent a triumph not only of engineering but of imagination, too. Removed from their mothers on Day 5 of gestation, when they were mere clusters of cells, these mice grew and thrived in their little vials for six days, growing tissue and organs—brain, blood, and tiny, pulsing hearts—until the experiment reached the limits of its viability on Day 11. While this new womb system, pioneered by cell biologists at Israel's Weizmann Institute of Science, may not yet have replicated a full-term mouse pregnancy, it's a mere whisker away: the usual gestation is a tantalizingly close nineteen days.[18,19]

Perhaps a more accurate vision of future womb technology is not the baby in the bag, or the jar on the shelf, but a spinning chamber, its apertures a metal honeycomb of human embryos perfectly calibrated to revolve to some unheard, vital rhythm. Perhaps the questions we ask are not so hypothetical, after all; perhaps we delude ourselves by thinking that our reproductive future holds divergent paths: one of flesh, one of steel. Perhaps there is no "if"; there is only "when."

—

WHO KNOWS WHAT CONCEPTION, PREGNANCY, AND BIRTH will look like in the future? Motherhood in the present has infinite iterations: glossy and filtered on social media; desperate but dignified in a prison cell or a migrant camp; fast and furious in a hospital car park; titrated by hormones and an epidural pump in the belly of the labor ward. In Jennifer Gobrecht's case, motherhood was a triumph to be featured in the centerfold of *People* magazine: there she is in the February 13, 2020, issue, all perfectly tonged hair and eyelashes, cradling her sleeping son, Ben, in her arms.[20] A soft-focus dream of maternal bliss. But all, as the saying goes, is not what it seems. Jennifer's pregnancy was not just any pregnancy. Ben is, quite literally, a miracle of modern science; his existence, the plot twist in a tale that some would find as gruesome as it is inspirational. On the day when

Jennifer and I "meet" by video, the gloss and the glamour have been stripped away, and the glowing rectangle of my screen shows only the truth, surprising and raw.

"Running a few minutes behind," Jennifer emails me as I wait for her at my desk. "Got stuck in a storm." It's Sunday evening for me in Scotland, but Jennifer is five hours behind me in Pennsylvania and she's just rushed home from a birthday party. When my screen flickers to life, here is a version of motherhood I recognize: Jennifer is harassed and apologetic, wet hair plastered across her forehead, face tense with the familiar anxiety of a woman trying to appear professional while keeping one ear open to the cries of a barely napping toddler in the next room. Jennifer and I have been exchanging polite emails for weeks ahead of this interview, but the summer downpour has swept any formality away. Our time is limited before Ben wakes up, and we get right down to it: how Jennifer was told she would never carry her own child in her body, and how she ended up doing exactly that. She didn't borrow a uterus from her sister or her mother, nor was she the first human recipient of a cutting-edge bio-womb. She carried Ben inside the kind of organ that's readily available right now, ready-made, fit for purpose; the kind of organ that's been tested and proved itself up to the task, but is buried or incinerated by the billion each year like so much human detritus. Jennifer smooths her hair across her face, I sit forward in my chair, and she tells me how she carried her child in the womb of a dead woman.

Ben's origin story begins on August 26, 2004, in a clinic that Jen describes in an Instagram post as "the place that used to be Bed Bath & Beyond on Baltimore Pike."[21] It was Jennifer's junior year of high school, and at age seventeen, she was the only one of her peers not to have gotten her period. A series of embarrassing and ultimately fruitless investigations had led to this women's health clinic in a nondescript building on the Pike, where Jennifer had come to discuss the results of a recent MRI. Jennifer remembers the day clearly, because it was her grandmother's birthday; she and her mother thought the appointment would be a quick pit stop before taking "Grandmom" out to lunch at Bennigan's.

"And that was the day they said, 'That's it, there's nothing there,'" Jennifer tells me. "'You're just born without a uterus. So good luck with that.'"

Like Dr. Brännström's transplant patient, Jennifer was diagnosed with MRKH syndrome, a condition that was completely alien to Jen, her mother, and even her doctor, who admitted having to look it up online. When asked if MRKH would make pregnancy impossible, the doctor replied with guarded optimism: "I never give someone a one hundred percent diagnosis. I believe there's always something new in science, so I'm going to give you a two percent chance to ever have your own child. You never know."

Lunch at Bennigan's may have been off the table, but for the next few years, Jennifer and her mother pursued that promising 2 percent with fierce determination. When she married her husband, Drew, in 2014, the couple initially considered adoption and surrogacy. Still, they continued to scan the news for any innovation that might give them another option.

"We started following the science," Jennifer recalls, "and right about that time, they started doing Swedish [uterus transplant] trials. We just thought it was so interesting."

By 2016, the Gobrechts had used IVF to create viable embryos using Drew's sperm and eggs from his wife's ovaries (which, like many women with MRKH, Jennifer still has). Just when it was beginning to look as though surrogacy would be the most practical pathway, Jennifer heard about a new transplant trial at the University of Pennsylvania—just minutes from her home—that was recruiting participants. A phone call put Jennifer on the list for interviews and assessments; the opportunity seemed too good to be true. If accepted, Jennifer wouldn't be getting an elaborate, expensive proto-womb machine; she would be offered the uterus of a deceased donor: a woman whose life had been cut short, but whose uterus had already borne a child, and still had the potential to do so again.

Harvesting organs from deceased donors is hardly new or unusual; tens of thousands of such procedures are carried out around the world each year, with kidneys, livers, hearts, and lungs being the most commonly donated organs in the UK and US.[22,23] Why not add the uterus to this list? Doing so would eliminate the considerable medical, psychological, and financial toll of uterus transplantation from live donors: the fraught dynamics of approaching a tissue-matched relative or friend for donation, the difficulty of selecting and safely operating on live donors, the grief when miscarriage ensues or the recipient's body rejects the donated organ, and the economic cost of a live donor's treatment and time away from the workplace. Even when one considers the possibility of artificial wombs, deceased donor wombs have the clear advantage of zero cost and ready availability; sadly, women in their reproductive prime die all the time for any number of reasons, most of which have nothing to do with the womb. Free, functional, and easy to find: it is hard to see why deceased uterus donation hasn't been the norm for years.

Or is it? Is growing your own embryo inside a dead woman's womb just a little bit . . . creepy? The media always loves a good organ donation story—the loving sisters who swapped kidneys, or the father who knows that his dead son's heart now beats in another young man's chest—but is it just too intimate, too ghoulish, to imagine that a woman like Jennifer could receive the womb of another woman and bear her own child within it? When I ask Jennifer if she had ever bristled at the idea, her reply was pragmatic and unequivocal.

"Why can't there be transplants that are life-enhancing, not just lifesaving? You can't take all this stuff with you when you die, so why not have someone else utilize it in a way that can truly enhance life; create life, even? It's such a miraculous organ, you know. Your kidney filters out toxins; this creates *people*."

Jennifer's clear-eyed view of the transplant process stood her in good stead during the rigorous psychological screening required by the UPenn team; this, and her physical fitness, earned her a place on the trial. And so it was that Jen found herself, like so many other would-be recipients whose fortunes depend on the demise of others, waiting for "The Call." In Jennifer's case, the fateful moment came on a Friday afternoon. The team had found a suitable donor: a twenty-nine-year-old woman who had already had children, and whose womb, therefore, had been tried and tested. In a letter Jennifer later received from the donor's own mother, the donor was described as "the best mother I ever knew."[24]

This young woman would never know the gift she gave to the Gobrechts. At the age of thirty-two, just a few months after the transplant, Jennifer had her first period. Finally, here was the rite of passage her friends had experienced so many years ago: the unmistakable sign of a functional uterus. Six months later, one of the IVF embryos was transferred successfully, followed by a pregnancy that Jennifer describes as completely straightforward and unremarkable apart from the near daily checkups and the dozens of immunosuppressant drugs required by the trial. A diagnosis of rapid-onset preeclampsia at thirty weeks meant that it was time for Baby Gobrecht to meet the outside world, and in what Jennifer describes as a "bittersweet" event, the trial team performed a Cesarean section followed by a hysterectomy. Keeping her womb beyond that pregnancy could have risked further, as yet unknown complications, and would have required Jennifer to continue taking antirejection medications—and enduring their often unpleasant side effects—indefinitely. In November 2017, Ben Gobrecht was delivered, and so, too, was the uterus that had nurtured him.

Just as "Angela" played her part in reproductive history by inspiring Mats Brännström, a new chapter has been written by Jennifer's donor and by the many other deceased donors whose uteruses have now been used in similar trials around the world. As immunologists, obstetricians, and specialist surgeons gain proficiency in such procedures, we will seek answers, too, to the ethical questions raised along the way. It may yet be the case that society's mixed feelings about the uterus—both reviled and revered—present a bigger obstacle than any technical hitch. But why? We accept that virtually every other body part, from kidneys to corneas, can be used in this way, so it seems irrational to attach some kind of holy, untouchable significance to the uterus. Over the last millennium, science has shown us that far from being "an animal within an animal" with its own fickle spirit, the uterus is an organ of neutral, functional character like any other. If the goal is to give as many women a chance at childbearing as possible, while sparing the health of potential live donors and taking advantage of a steady supply of cadavers, then why should this next step be any more controversial than the last?

Certainly, early discourse around uterus transplants was dominated by the idea of the womb as an organ with special status. In their 2007 article "Moving the Womb," bioethicist Arthur Caplan and his coauthors suggest that uterus transplants may be an emotional step too far for women who would otherwise have been in favor of organ donation: "Few, if any, American women ever thought that the uterus might be one of the organs considered for donation when they signed a donor card. A woman might not prove as willing to donate her uterus as she would be to donate her heart or her liver."[25] The rationale behind this reluctance is not explored in great detail; rather, it seems to be implicitly understood that a woman's gut feeling about her uterus—the notion that it is uniquely and irreversibly *hers*, more than any other organ—is reason enough. More than a decade after Caplan's article was published, this theory has yet to be tested in any systematic, rigorous way: an essential prerequisite to the establishment of any large-scale uterus donation program. For the time being, there is the small but steadily growing number of women who have participated in clinical trials as living uterus donors, and there are the families who have opted to donate the womb of a deceased loved one, too. As the mother of Jennifer Gobrecht's donor also wrote in her letter, "What a beautiful and fitting legacy for [my daughter] to help give the gift of motherhood to another woman."[26]

Jennifer says that while the public response to her story is generally positive, she's familiar with what she calls "the backlash," too. In her experience, this opposition to uterus transplants has less to do with any reluctance from potential donors or their families, and more to do with the perceived selfish or unreasonable demands of the recipients.

"There are always the people who are like, 'I just don't understand why you'd want to go through such an extreme fertility treatment; just go adopt.' Like there's some store I could just go to, like that's an easy mental thing," she says. "People who just aren't in it don't understand why someone would want to experience pregnancy and childbirth. And you know, different people have different preferences of what they want to experience in life." In this kind of ideological debate, there can be no clear winner. As Sheila Heti writes in *Motherhood*—an auto-fictional exploration of choices around childbearing—"A woman will always be made to feel like a criminal, whatever choices she makes, however hard she tries. Mothers feel like criminals. Non-mothers do, too."[27] One person's decision to have a child—

and their chosen method of doing so—will inevitably be anathema to another person facing the same fork in the biological road.

Ultimately, Jennifer says, technologies like uterus transplants—whether from live or deceased donors—are about providing choice, something that is in scant supply for many women in their childbearing years.

"I hope that women with different infertility conditions look at this as another path that they can take, because it's very hard when you're a young woman to be told that you don't have many options. I think the more options we're able to create for women is really important. That's why I wanted to be one of the first people to try this, because even if it didn't work for me, we can at least find more ways to make it better for other people."

It's worth noting, too, that this group of "other people" may be even more inclusive in the future: scientists have already begun to explore the possibility of transplanting wombs into bodies with physiologically male characteristics. Specifically, they theorize that trans women—people who were assigned male at birth, and may still retain some or all of their male anatomy while identifying and living as females—could be uterus recipients if and when science and society allow. Authors of a 2019 review concluded, "Despite a number of anatomical, hormonal, fertility and obstetric considerations . . . there is no overwhelming clinical argument against performing UTx [uterus transplant] as part of GRS [gender reassignment surgery]." The authors go even further in their argument for reproductive equity, suggesting that if these medical obstacles can be safely overcome, it may in fact be immoral to exclude this population from the possibility of gestating and birthing children: "The reproductive aspirations of M2F [male to female] transgender women deserve equal consideration to those assigned female at birth and, subject to feasibility being shown in the suggested areas of research, it may be legally and ethically impermissible not to consider performing UTx in this popula-tion."[28] This hypothetical scenario may not have been envisaged by the family of Jennifer's donor—or even by "Angela," who inspired Mats Brännström's earliest innovations—but it may become a reality within our lifetimes.

It's getting late for me in Scotland, and the end of naptime is approaching for Jennifer in Pennsylvania; I can see her eyes dart-ing with increasing frequency to the sleeping child off-screen. As we say our goodbyes and return to our respective domestic obligations, I wonder whether giving options to people—women, men, same-sex and nonbinary couples, anyone with a uterus and anyone who wants one—can be such a bad thing? Choice—about if, when, and how to have sex, start a family, or end a pregnancy—enables women and people with wombs to lead safe, satisfying lives. In turn, making those choices allows these people not just to survive in our world but to thrive in it, becoming fulfilled individuals as well as engaged, em-powered members of their communities. Reproductive choice—like food, water, and shelter—is not a luxury, but a necessity.

—

FOR ANYONE WITH A WOMB, THE UTERUS IS THE NEXUS where the ley lines of reproductive choice converge. Nowhere is this convergence more evident or more contentious than in the domain of abortion. Every government in the world has attempted to deter-mine if, how, and when an individual should have the legal right to terminate a pregnancy. There is no global consensus about when life begins, or whether that life is more valuable than that of the person gestating it. There is no universal acceptance of the notion that a person's body—or at least, the womb within their body—is their own sovereign territory. Regardless of this philosophical sticking point, abortion and abortifacient substances have—as we've seen—been omnipresent throughout history, and the procedure itself continues to be an essential component of reproductive healthcare. In some cases, termination of a pregnancy can avert life-threatening medical crises; in others, it can have less immediately obvious but no less beneficial effects on a person's physical, emotional, and even financial well-being.[29]

The World Health Organization's stance on abortion is unequivocal: "Abortion is safe when carried out using a method recommended by WHO, appropriate to the pregnancy duration and by someone with the necessary skills."[30] In spite of this endorsement, many countries continue to impose restrictive laws or outright bans on abortion. Such draconian legislation of the uterus and its contents hardly prevents unwanted pregnancies or stems the need for abortion. On the contrary, it drives people to seek help from unskilled providers in unhygienic environments, with the tragic—and some might say, inevitable—consequence that tens of thousands of women die each year from complications of unsafe abortions.[31] Suffering from infection, hemorrhage, and damage to the uterus and surrounding organs,[32] these women are quite literally dying to take control of their wombs.

While 97 percent of unsafe abortions occur in developing countries,[33] access to this lifesaving procedure is not only a Third World problem: pregnant people in some of the world's wealthiest, so-called progressive countries are dying, too, from the consequences of laws that prioritize the life of the fetus over that of the mother. The faces of these women appear with haunting regularity on front pages and in news bulletins, and their stories are disturbingly similar. In 2012, global media turned its attention to the story of Savita Halappanavar, who sought help in an Irish hospital for what appeared to be the imminent loss of a seventeen-week pregnancy. Despite the fact that Savita became progressively sicker with sepsis secondary to an infection within the womb, doctors could not legally induce termination of the pregnancy as long as the fetus's heart continued to beat. By the time Savita spontaneously miscarried, the ongoing pregnancy had caused such severe infection that septic shock, multiple organ failure, and cardiac arrest soon followed.[34] The widespread coverage of Savita's death on October 28, 2012, undoubtedly played a role in the subsequent public vote to repeal Ireland's restrictive abortion law, but that change came too late for Savita herself. Since her death, eerily similar narratives have unfolded in the cases of other women for whom lifesaving abortion came too late, or not at all. In early 2022, a Polish woman known publicly as "Agnieszka T" died just over a month after presenting to the hospital with the threatened miscarriage of a first-trimester twin pregnancy; Agnieszka's family allege that her death from septic shock was caused by the hospital's delay in inducing delivery of the twins, even though both fetuses had died in the womb within days of Agnieszka's admission.[35] As with the case of Savita Halappanavar and so many like her, the flicker of a fetal heartbeat—even one that was likely to cease spontaneously—appears to have been given greater priority than the life of that fetus's mother. Uterine infections can take hold with alarming strength and speed, especially after fetal demise. In these cases, timely evacuation of the uterus by medical or surgical means is critical: a fact which appears to hold little weight in countries with the most punitive abortion laws.

Perhaps the most chillingly retrograde legislation of the uterus can be found not in Ireland or Poland, but in one of the world's most politically and socially powerful countries: America. Since Donald Trump jibed that Megyn Kelly was bleeding from her "wherever," a wave of restrictive abortion laws—some of them outright bans in all but name—has swept across the United States. In May 2021, Texas governor Greg Abbott signed SB8, the so-called heartbeat law, which prohibits abortion as soon as fetal cardiac activity can be detected.[36] According to SB8, this occurs at six weeks of pregnancy; in other words, six weeks after the first day of a person's last menstrual period, and for many people, before they may even realize they're pregnant, let alone have time to choose and seek an abortion. Many states followed suit with extreme curbs on the timing and circumstances of legal abortion; more laws like these were passed by state legislatures in 2021 than in any year since 1973, when the landmark case of *Roe v. Wade* enshrined the right to abortion in America.[37] By early 2022, three states had attempted to ban abortion throughout pregnancy, eight had attempted to ban abortion at six weeks,[38] and twelve states had so-called trigger laws that would ban all or almost all abortions in the event of *Roe v. Wade* being repealed.[39]

On June 24, 2022, the gathering storm on the horizon of America's reproductive rights exploded in a deafening thunderclap. The US Supreme Court, a bench increasingly stacked with right-leaning judges in recent years, announced its ruling in the case of *Dobbs v. Jackson Women's Health Organization*. Originally a challenge to restrictive abortion laws in the state of Mississippi, *Dobbs*'s escalation to the Supreme Court represented the most significant legal challenge yet to Americans' bodily autonomy. In his written decision summarizing the Court's 6–3 majority ruling, Associate Justice Samuel Alito makes an eclectic and undoubtedly selective survey of European and American history, citing what he perceives as a long tradition of anti-abortion legislation and ideology dating back to medieval times. He arrives—perhaps inevitably—at the conclusion that the right to abortion is not, and never was, protected by the Constitution. "Roe," he writes, "was egregiously wrong from the start."[40]

With one stroke of his pen, Alito dealt a stinging blow to pregnant people and their care providers. From now on, in many parts of the country, an American with a fetus in her womb no longer has control over her own body. Regardless of factors such as age, gestation, social needs, risk to the mother's life, or dire circumstances including rape or incest, she may be forced to carry that fetus to term. Each state may determine whether abortion is legal within its boundaries, and each state may impose its own punishments—heavy fines, felony charges, and/or jail time—on those who seek, aid, or abet the procedure.

Although this ruling has set off shockwaves that will, no doubt, reverberate for years to come, perhaps we should not be surprised. The ink on Alito's decision may barely be dry, but his writing, so to speak, has been on the walls for years. Antichoice rhetoric has been increasingly prominent in American culture in the late twentieth and early twenty-first centuries—espoused not only in online forums and on protest placards but also displayed on objects as mundane as the family car. "Choose Life" license plates—some of which bear images of smiling infants and baby footprints, and many of which raise funds for anti-choice organizations—are available in thirty-three states.[41] In contrast, can one imagine a world in which a license plate is used to endorse and fund the restriction of men's reproductive rights? Can one envision a license plate that advertises, say, forced vasectomies, accompanied by an image of shiny scissors or a surgeon's knife? One cannot. The comparison may seem absurd, but the inequity is stark and pervasive. It was all around Americans, all the time: in private conversations and public debates, and on a vast convoy of vehicles cruising smoothly toward social cataclysm.

At this tenuous time, the author Margaret Atwood invites us to interrogate our values, as well as what those values might mean for the future of reproductive autonomy: "We have to ask: what kind of country do we want to live in? A democratic one in which every individual is free to make decisions concerning their health and body, or one in which half the population is free and the state corrals the bodies of the other half?"[42] This question is not unique to one person or one state; the "we" to whom Atwood refers includes all of us, in every part of the world. For many people with a womb, though, freedom of choice remains elusive. *My Body Is My Own*, a 2021 report from the UNFPA (the United Nations' sexual and reproductive health agency), presents the results of a survey in which women from fifty-seven countries were asked about their bodily autonomy. Only 55 percent of respondents said they were able to make their own decisions about sexual and reproductive health and rights; this included whether they were able to say no to sex, use contraception, or access healthcare. Conversely, 45 percent—nearly half the women in the survey—felt otherwise.[43] For this not-insignificant minority, autonomy comes a distant second to the physical, sexual, and reproductive demands of patriarchal society and its agents, from the halls of government to the heart of the home. From America, where new threats to reproductive choice emerge on an almost weekly basis, to the twenty countries—including Bahrain, Bolivia, the Russian Federation, and the Philippines—where rape convictions can be overturned if the perpetrator marries his victim,[44] women's bodies and their wombs are not their own. Not wholly, and therefore, one could argue, not at all. To be partially or conditionally autonomous—free, but only insofar as it suits another's needs—is an irreparable loss and a fundamental injustice.

Even in the most forward-thinking countries, we may choose which foods to put in our bellies, which tools to grasp with our hands, which private thoughts to harbor in our brains, but for so many of us—in the past, the present and, undoubtedly, far into the future—we may not always choose what to do with our wombs, or how to pursue or avoid parenthood according to our hopes and desires. We lack this fundamental freedom, and all too often, we lack the knowledge of the uterus's basic functions: how it grows, bleeds, births, and transforms with life's ever-changing tides. Many of us lack even the simplest words to describe the womb and what it does, and where there is no language, there can be no self-expression.

This book is simply that—language—but these dry dots and dashes on a page speak of an organ that is vital, blood-flushed, and pulsing with life. An organ that is linked inextricably to our biological, social, and political destinies. An organ that would tell its story—tragic, triumphant, and endlessly evolving—if only we would listen.

An Unapologetic Epilogue;
or, An Invitation
to the Reader

A writer friend told me recently that when female authors come to the end of a work in progress, they almost always write an epilogue whose sole purpose is to serve as an apology for the many perceived flaws of the preceding pages. Her words sparked a pang of self-recognition. I hadn't realized until our conversation that that was exactly what I had been planning to do.

Throughout the writing of this book, I've been making mental notes of all the things I could say in my conclusion that would in some way mitigate or excuse the limitations of my work. While it is true that this book may have its unintended omissions and even (although I have done my very best to prevent them) inaccuracies, to close this work by highlighting these flaws would be counterproductive, and contrary to the intentions of the book itself. I began with the aim of understanding—and perhaps even celebrating—an oft-maligned and neglected organ, and the bodies and lives of those who have one. So rather than indulge in pointless (and, according to my friend, drearily predictable) self-flagellation, I will turn with purpose against my weaker instincts and, instead, reflect on the lessons I have learned from those interviewees and experts who have guided me so generously on this journey.

At the outset, gazing in wonder at those disembodied wombs at the Surgeons' Hall Museums, I thought I knew exactly what I would write and how I would write it, the structure and intention of my book as clear as the formaldehyde in the jars in front of me. As I began to research and write, though, the uterus, its history, and its enthusiasts showed me that I needed to discard my preconceived ideas in order to develop an understanding that was holistic, accurate, and forward-thinking. What I realized quite early on in this process is that any understanding of the uterus must be as cyclical as the functions of the organ itself. No single part of the womb's existence can be viewed in isolation, pinned to the literary foreground like a butterfly under glass. Rather, each aspect of the uterus and its significance within our lives is linked to the next, and the next, and loops back to the previous, and so on ad infinitum. One cannot understand the adult womb without first exploring the organ in its infancy. One cannot discuss menstruation without looking backward to childhood, and forward to menopause. One cannot appreciate the strong contractions of a uterus in labor without first understanding the minute but no less important pulses of the womb in the first moments of conception, and one cannot fully appreciate the joy of birth without stopping awhile in the dark shadows cast by pregnancy loss. It may seem a bit airy-fairy—a bit "woo"—to argue that a book about the uterus must organically follow the rhythms and loops of the reproductive cycle itself, but this is the way the womb has led me since this project's earliest beginnings. I hope you, too, have been encouraged by these pages to expand your knowledge, join it up, double it back on itself, and integrate it into a new and more exciting whole, one that reflects the seasons of our lives with authenticity and respect.

Not every lesson I learned in the writing of this book has been as easy or as palatable to swallow. It quickly became apparent to me that, as much as any understanding of the uterus must be cyclical, it must be intersectional, too. One cannot generalize the experience of having a womb across the entire population; each person's uterine life will be deeply colored by the overlapping shades and tones of their identity. Whatever troubles a uterus might bring, those troubles are often harder, more painful, and more frequently dismissed—or even held in contempt—by medicine, law, and society at large if you happen to be Black or Brown. And if you belong to a group that is in any way marginalized, "othered," or dominated—if, say, you are poor, enslaved, incarcerated, or migrant; or if you do not fit the gender norms of the culture around you—then those troubles are more onerous still. This is not some kind of virtue-signaling wokery on my part. This is fact, borne out by scientific and sociological evidence. As a cisgender, heterosexual, educated white woman with a steady job and secure citizenship, it would be disingenuous of me to claim these struggles as my own, but what I have tried to do is to amplify the voices of those for whom the uterus is a painful matrix of oppression; those whose fertility, sexuality, identity, and health have been demeaned and even, in some cases, destroyed. I hope this book has channeled those voices—some loud and indignant, some soft yet persistent—onto the page and into your ear. Some of these voices may echo yours; you may already have navigated a similar landscape of injustice and inequity. Conversely, you may not feel as though period poverty, coercive sterilization, and medical racism are part of your world, but please know that they are, and know that owning a uterus carries a heavy price for those people affected by these issues. Honor their experiences by reading, listening, and reflecting on whether you are part of the problem, or the solution, or—as many of us are likely to be—somewhere in the complex gray area in between.

Advancing our collective understanding of the uterus and its impact on all our lives is largely dependent on the willingness of government bodies to fund research into these areas. Sadly, the budgets of those organizations are often controlled by people—quite often men—who view this kind of work as relatively unimportant, of "niche" interest to a precious few, or unprofitable. Recall how Christine Metz told me that her literature search of scientific publications yielded many more results for "semen" than for "menstrual blood"—fifteen thousand versus a meager four hundred. These results could be seen as representative of science's (and the wider world's) disproportionate interest in male bodies—their function, their health, and their pleasure. This discrepancy seems shortsighted at best and dangerous at worst, considering that roughly half of the world's population is born with a uterus, and all of us began our life's journey inside one.

Inequities in funding for women's health have only been exacerbated by the Covid-19 pandemic. In a state of physiological extremis, the body diverts its blood supply toward the most essential organs—the heart and the lungs—and away from those less intrinsic to immediate survival. This rationing of resources has been mirrored on a global scale during the pandemic. As governments have plowed money and (wo)manpower into prevention and treatment of the virus, other areas of care—such as sexual health, obstetrics, and gynecology—have languished. The uterus and its owners have suffered disproportionately. In some places, women have struggled to access contraception and abortion; supply issues have left women without their usual hormone therapy for menopause; pregnant people have experienced service cutbacks and distressing restrictions on partners in maternity hospitals; and midwives have faced unprecedented levels of stress and burnout while caring for critically ill patients within an already underfunded, understaffed system. At the heart of each of these issues is the womb, going about its business of bleeding and breeding as diligently as ever, oblivious to global events. Its needs are as acute as they've always been. Waves of infection may ebb and flow, viral spikes may rise and fall, but the uterus continues to strive and suffer amid this flux. Epidemiologists insist that another pandemic is inevitable; will the womb fare any better the next time around? Or, in the rush to protect and preserve life, will the most vital of organs—the one so necessary for the very survival of our species, virus or no virus—be sidelined once again?

I'm under no illusion that this book will make the holders of the world's scientific purse strings suddenly loosen their grip, at this or any other time in the future. Funding for research like the ROSE trial, or for Margherita Turco's work on endometrial organoids, or for Monica Tolofari and Linn Shepherd's campaign for the safe use of oxytocics, is and will continue to be hard to come by: like drawing blood from the proverbial stone. What we can all do now—at no cost to ourselves or others, or to the financial powers-that-be—is to learn more about ourselves. How does having a womb affect your life? What language do you use to describe or disparage that organ? Has it brought you pleasure or pain, or perhaps a complex web of both, as tightly woven as the fibers of the uterus itself? Do you understand its functions and its dysfunctions, its phases and its cycles, from month to month, from birth to death?

As I wrapped up my interview with Rebecca Fischbein, she expressed a humbling gratitude for the fact that such a dangerous and frightening experience of twin-to-twin transfusion had prompted a deeper understanding of and appreciation for her body. I remarked that so many of the women I had spoken to had taken a similar trajectory: a personal trauma leading to a lifelong quest for knowledge, and sometimes even translating into a new professional path. "Yeah," Rebecca said, nodding in recognition of this common phenomenon. "My friend and I call it *me-search*."

I will close this book, then, not with an apology, but with an invitation: if you have a womb, or if you live with or care about someone who does, or even if you haven't given the uterus much thought since you emerged from one, blood-slicked and screaming, many years ago, do your own me-search. Interrogate and celebrate your experience. Understand that fist-shaped muscle, that powerful source, that place where we all began. It might even tell us, in so many ways, where we are going.

ACKNOWLEDGMENTS

This book started with a very tentative email to my agent, and ended with a community of experts, enthusiasts, and supporters beyond my wildest dreams. I am so grateful to each and every person who has guided *Womb* through its gestation. The birth was laborious, but the midwives were excellent.

About that agent: Hayley Steed is the best one. Thank you, Hayley, for believing in me ever since our first meeting as publishing rookies. Let's definitely keep doing this. Thanks also to Liane-Louise Smith for convincing Hayley that *Womb* was a good idea, and for pitching that idea around the world along with the wonderful Georgina Simmonds.

For fierce advocacy, keen editorial skill, and kind indulgence of this author's neuroses, I thank Rose Tomaszewska at Virago and Sara Birmingham at Ecco. Thank you, too, to Denise Oswald for early support, Zoe Carroll for taking the reins, Mary Chamberlain for the most immaculate copyediting (again), and Alison Griffiths for help at the final furlong. *Gracias* to David Orión Pena Carpio for ensuring that this book is as inclusive and compassionate as possible. Thank you—in some languages I know, and in many I don't—to my international publishers and translators.

Thanks to Lee Randall for seeing a bestseller in my one-line pitch and to Jane Healey for listening to my woes on writerly walks. Mary Renfrew and Sue Macdonald, midwifery legends: I owe you a debt of gratitude, and you have my admiration.

Womb would not exist without the many contributors and

interviewees who shared their expertise and lived experiences, often amid varying degrees of lockdown-induced domestic chaos. I cannot thank you enough for your generosity of time, wisdom, and spirit. Thank you also to those who provided valuable context, including Marilen Benner, Louise Wilkie, and Sebastian Hofbauer.

As I wrote, thousands of voices echoed in my ear: the voices of the women and families I've had the privilege of supporting in my career as a midwife. Their words may not have made it to the page, but their strength, wit, and dignity reverberate through this book. They are my coauthors, always.

Other voices coaxed, cheered, guided, and chided me in my "day job" throughout the writing of *Womb*. These are the voices of my teachers and friends: my fellow midwives, as well as the doctors, auxiliaries, porters, and associated staff who keep the maternity machine running. Thank you for bearing with me and for keeping my feet firmly on the ground. My gratitude, too, to the many midwives and student midwives who have supported me on social media.

Nobody knows the journey of an author better than those who view it firsthand. Thank you to my family in America and Scotland, with special mention to my father, who didn't live to see *Womb*'s publication but who, after reading a rough draft, insisted that "everyone should buy this."

Love—all of it, forever—to A., S., and A. It's really for you.

Introduction: In Search of the Womb

1. *Vagina Dialogues* press release, Eve Appeal, July 2016, eveappeal.org.uk/wp-content/uploads/2016/07/The-Eve-Appeal-Vagina-Dialogues.pdf.
2. H. Scott, "Half of Men Don't Know Where Vagina Is, According to a New Survey," *Metro*, August 31, 2017.
3. A. Y. Sherwani et al., "Hysterectomy in a Male? A Rare Case Report," *International Journal of Surgery Case Reports* 5, no. 12 (2014): 1285–7.
4. C. Pleasance, "Businessman to Have a Hysterectomy After Discovering He Has a WOMB as Well as Normal Male Organs," MailOnline, February 9, 2015, dailymail.co.uk/news/article-2952983/Pictured-time-British-businessman-set-hysterectomy-discovering-WOMB-normal-male-organs.html.

Uterus

1. H. J. Paltiel and A. Phelps, "US of the Pediatric Female Pelvis," *Radiology* 270, no. 3 (March 2014): 644–57.
2. T. Escherich, "The Intestinal Bacteria of Neonates and Their Relationship to the Physiology of Digestion," thesis published in 1886, cited in J. Hacker and G. Blum-Oehler, "In Appreciation of Theodor Escherich," *Nature Reviews Microbiology* 5 (2007): 902.
3. H. Tissier, "Recherches sur la flore intestinale des nourrissons (état normal et pathologique)," Paris: G. Carre and C. Naud, 1900, cited in A. A. Kuperman and O. Koren, "Antibiotic Use During Pregnancy: How Bad Is It?" *BMC Medicine* 14, no. 91 (June 2016).
4. D. Parton, "These Old Bones," Velvet Apple Music, 2002.
5. E. Jiménez et al., "Is Meconium from Healthy Newborns Actually Sterile?" *Research in Microbiology* 159, no. 3 (2008): 187–93.
6. L. F. Stinson et al., "The Not-So-Sterile Womb: Evidence That the Human Fetus Is Exposed to Bacteria Prior to Birth," *Frontiers in Microbiology* 10 (2019): 1124.
7. M. Benner et al., "How Uterine Microbiota Might Be Responsible for a Receptive, Fertile Endometrium," *Human Reproduction Update* 24, no. 4 (July–August 2018): 393–415.
8. M. E. Perez-Muñoz et al., "A Critical Assessment of the 'Sterile Womb' and 'In Utero Colonization' Hypotheses: Implications for Research on the Pioneer Infant Microbiome," *Microbiome* 5, no. 48 (2017).
9. H. Verstraelen et al., "Characterisation of the Human Uterine Microbiome in Non-Pregnant Women Through Deep Sequencing of the V1-2 Region of the 16S rRNA Gene," *PeerJ* 4 (January 19, 2016): e1602.

10. S. Dizzell et al., "Protective Effect of Probiotic Bacteria and Estrogen in Preventing HIV-1-Mediated Impairment of Epithelial Barrier Integrity in Female Genital Tract," *Cells* 8, no. 10 (2019): 1120.

11. P. Moayyedi et al., "Fecal Microbiota Transplantation Induces Remission in Patients with Active Ulcerative Colitis in a Randomized Controlled Trial," *Gastroenterology* 149, no. 1 (2015): 102–109.

12. R. Tariq et al., "Efficacy of Fecal Microbiota Transplantation for Recurrent *C. Difficile* Infection in Inflammatory Bowel Disease," *Inflammatory Bowel Diseases* 26, no. 9 (September 2020): 1415–20.

13. International Clinical Trials Research Platform Search Portal, World Health Organization website, accessed November 30, 2021, who.int/clinical-trials-registry-platform.

14. Benner et al, "How Uterine Microbiota Might Be Responsible for a Receptive, Fertile Endometrium."

15. N. K. Dinsdale et al., "Comparison of the Genital Microbiomes of Pregnant Aboriginal and Non-aboriginal Women," *Frontiers in Cell and Infection Microbiology* 10 (October 29, 2020): 523764.

16. N. M. Molina et al., "New Opportunities for Endometrial Health by Modifying Uterine Microbial Composition: Present or Future?" *Biomolecules* 10, no. 4 (April 2020).

Periods

1. I. S. Fraser et al., "Blood and Total Fluid Content of Menstrual Discharge," *Obstetrics and Gynecology* 65, no. 2 (1985): 194–8.

2. Cambridge Dictionary online, dictionary.cambridge.org/dictionary/english/effluent.

3. E. Martin, "The Egg and the Sperm: How Science Has Constructed a Romance Based on Stereotypical Male–Female Roles," *Signs* 16, no. 3 (1991): 485–501.

4. A. Nayyar et al., "Menstrual Effluent Provides a Novel Diagnostic Window on the Pathogenesis of Endometriosis," *Frontiers in Reproductive Health* 2, no. 3 (2020).

5. S. Toksvig, "And Woman Created . . . ," *Guardian*, January 23, 2004.

6. J. Abbink, "Menstrual Synchrony Claims Among Suri Girls (Southwest Ethiopia): Between Culture and Biology," *Cahiers d'Études Africaines* 55, no. 2018 (2015): 279–302.

7. A. H. Gupta and N. Singer, "Your App Knows You Got Your Period. Guess Who It Told?" *New York Times*, January 28, 2021.

8. A. Bhimani, "Period-Tracking Apps: How Femtech Creates Value for Users and Platforms," *LSE Business Review*, May 4, 2020.

9. S. Dunn, online message to author, February 10, 2021.

10. Bhimani, "Period-Tracking Apps."

11. C. Healy, online message to author, February 10, 2021.

12. Gupta and Singer, "Your App Knows You Got Your Period."

13. Clue, Twitter post, February 18, 2021, twitter.com/clue/status/1362342890152873990.

14. R. Hadley et al., "Use of Menstruation and Fertility App Trackers: A Scoping Review of the Evidence," *BMJ Sexual and Reproductive Health* 47, no. 2 (April 2020).

15. L. Hampson, "Women Spend £5,000 on Period Products in Their Lifetime," *London Evening Standard*, November 28, 2019.

16. O. Petter, "Period Pains Responsible for Five Million Sick Days in the UK Each Year," *Independent*, October 14, 2017.

17. M. Renault, "Why Menstruate If You Don't Have To?" *The Atlantic*, July 17, 2020.

18. S. Walker, "Contraception: The Way You Take the Pill Has More to Do with the Pope Than Your Health," *The Conversation*, January 22, 2019, theconversation.com /contraception-the-way-you-take-the-pill-has-more-to-do-with-the-pope-than-your -health-109392.

19. K. A. Hasson, "Not a 'Real' Period? Social and Material Constructions," in C. Bobel et al., eds., *Palgrave Handbook of Critical Menstruation Studies* (London: Palgrave, 2020), 7.

20. A. Edelman et al., "Continuous or Extended Cycle vs. Cyclic Use of Combined Hormonal Contraceptives for Contraception," *Cochrane Database Systematic Review*, July 29, 2014.

21. FSRH press release, January 21, 2019, Faculty of Sexual and Reproductive Healthcare, fsrh.org/news/fsrh-release-updated-guidance-combined-hormonal-contraception/.

22. H. K. Bradshaw, S. Mengelkoch, and S. E. Hill, "Hormonal Contraceptive Use Predicts Decreased Perseverance and Therefore Performance on Some Simple and Challenging Cognitive Tasks," *Hormones and Behavior* 119 (March 2020): 104652.

23. FSRH Guideline: Combined Hormonal Contraception, January 2019 (Amended November 2020). Faculty of Sexual and Reproductive Healthcare, fsrh.org/standards -and-guidance/documents/combined-hormonal-contraception.

24. C. S. Hopkins and T. Fasolino, "Menstrual Suppression in Girls with Disabilities," *Journal of the American Association of Nurse Practitioners* 33, no. 10 (October 2021): 785–90.

25. Y. A. Kirkham et al., "Trends in Menstrual Concerns and Suppression in Adolescents with Developmental Disabilities," *Journal of Adolescent Health* 53, no. 3 (2013): 407–12.

26. crippledscholar blog, July 8, 2016, crippledscholar.com/2016/07/08/lets-talk-about-disability-periods-and-alternative-menstrual-products/.

27. J. Wilbur et al., "Systematic Review of Menstrual Hygiene Management Requirements, Its Barriers and Strategies for Disabled People," *PLOS ONE* 14, no. 2 (2019): e0210974.

28. H. O. D. Critchley et al., "Menstruation: Science and Society," *American Journal of Obstetrics and Gynecology* 223, no. 5 (November 1, 2020): 624–64.

Conception

1. N. Ephron, *When Harry Met Sally*, Columbia Pictures, 1989.

2. J. Singer and I. Singer, "Types of Female Orgasm," *Journal of Sex Research* 8, no. 4 (1972): 255–67.

3. C. M. Meston et al., "Women's Orgasm," *Annual Review of Sex Research* 15 (2004): 173–257.

4. Obituary of Irving Singer, MIT News, February 8, 2015, news.mit.edu/2015/irving-singer-obituary-0208.

5. Obituary of Josephine (Fisk) Singer, Robert J. Lawler & Crosby Funeral Home, October 1, 2014, currentobituary.com/obit/146061.

6. E. Matsliah, "There Are 8 Kinds of Female Orgasms—Here's How to Have Them All!" YourTango, May 26, 2021, yourtango.com/experts/eyal-intimatepower/8 -different-female-anatomy-orgasms-and-how-reach-them.

7. "All About Orgasms: Why We Have Them, Why We Don't, and How to Increase Pleasure," Our Bodies, Ourselves online, October 15, 2011 (updated September 12, 2014), ourbodiesourselves.org/book-excerpts/health-article/all-about-orgasms/.

8. B. R. Komisaruk et al., "Women's Clitoris, Vagina, and Cervix Mapped on the Sensory Cortex: fMRI Evidence," *Journal of Sexual Medicine* 8, no. 10 (2011): 2822–30.

9. M. Roach, *Bonk* (Edinburgh: Canongate, 2009), 87–108.

10. L. Wildt et al., "Sperm Transport in the Human Female Genital Tract and Its Modulation by Oxytocin as Assessed by Hysterosalpingoscintigraphy, Hysterotonography, Electrohysterography and Doppler Sonography," *Human Reproduction Update* 4, no. 5 (September 1998): 655–66.

11. Instituto Bernabeu, September 9, 2020, institutobernabeu.com/en/news/instituto -bernabeu-study-relates-progesterone-to-uterine-contractility-and-its-effect-on -patients-with-embryo-implantation-failure/.

12. B. Moliner, email message to author, April 8, 2021.

13. E. Martin, "The Egg and the Sperm: How Science Has Constructed a Romance Based on Stereotypical Male-Female Roles," *Signs* 16, no. 3 (Spring 1991): 485–501, quoted in R. Martin, "The Idea That Sperm Race to the Egg Is Just Another Macho Myth," Aeon Essays, August 23, 2018, aeon.co/essays/the-idea-that-sperm-race-to -the-egg-is-just-another-macho-myth.

14. G. Bettendorf, "Insler, Vaclav," in G. Bettendorf, ed., *Zur Geschichte der Endokrinologie und Reproduktionsmedizin* (Berlin: Springer, 1995).

15. V. Insler et al., "Sperm Storage in the Human Cervix: A Quantitative Study," *Fertility and Sterility* 33, no. 3 (1980): 288–93.

16. "Sperm Trapped in Cervical Crypt," posted by D. Barlow, May 9, 2015, YouTube, youtube.com/watch?v=ho5u5MapiLs.

17. Bettendorf, "Insler, Vaclav."

18. S. Rhimes and P. Nowalk, *Grey's Anatomy*, season 7, episode 4, first aired October 14, 2010.

19. C. Goerner, "They Said I Have a Hostile Uterus—I'm Sorry, What?!" Bolde, bolde .com/hostile-uterus-sorry-what.

Pregnancy

1. M. Y. Turco et al., "Trophoblast Organoids as a Model for Maternal–Fetal Interactions During Human Placentation," *Nature* 564 (2018): 263–67.

2. M. Y. Turco et al., "Long-Term, Hormone-Responsive Organoid Cultures of Human Endometrium in a Chemically Defined Medium," *Nature Cell Biology* 19, no. 5 (2017): 568–77.

3. G. Berkers et al., "Rectal Organoids Enable Personalized Treatment of Cystic Fibrosis," *Cell Reports* 26, no. 7 (2019): 1701–8.

Tightenings

1. D. M. Fraser and M. A. Cooper, eds., *Myles Textbook for Midwives*, 15th ed. (London: Elsevier, 2009).

2. P. Dunn, "John Braxton Hicks (1823–97) and Painless Uterine Contractions," *Archives of Disease in Childhood. Fetal and Neonatal Edition* 81 (1999): F157–8.

3. Ibid.

4. J. B. Hicks, "On the Contractions of the Uterus Throughout Pregnancy: Their Physiological Effects and Their Value in the Diagnosis of Pregnancy," *Transactions of the Obstetrical Society of London* 13 (1871): 216–31.

5. Ibid.
6. "Robert Gooch," Royal College of Physicians Museum, history.rcplondon.ac.uk /inspiring-physicians/robert-gooch.
7. Ibid.
8. J. S. Coghill, *Glasgow Medical Journal* 7, no. 26 (1859): 177–86.
9. Ibid.
10. F. W. Mackenzie, "On Irritable Uterus," *London Journal of Medicine* (May 1851): 385–401.
11. Ibid.
12. Ibid.
13. Ibid.
14. R. Ferguson, ed., "Gooch on Some of the Most Important Diseases Peculiar to Women: With Other Papers," *New Sydenham Society* 2 (1859).
15. Ibid.
16. ICD10Data website, 2021, icd10data.com/ICD10CM/Codes/O00-O9A/O60 -O77/O62-/O62.2#:~:text=12–55%20years)-,O62.,ICD-10-CM%20O62.
17. R. Fischbein, "The Irritable Uterus," in A. Perzynski, S. Shick, and I. Adebambo, eds., *Health Disparities: Weaving a New Understanding Through Case Narratives* (Cham, Switzerland: Springer, 2019), 41–42.

Labor

1. Name changed for confidentiality.
2. "NHS Maternity Statistics, England, 2020–21," NHS Digital, digital.nhs.uk/data -and-information/publications/statistical/nhs-maternity-statistics/2020-21.
3. "Natality Statistics 2016–2020," Centers for Disease Control and Prevention, wonder .cdc.gov/controller/datarequest/D149;jsessionid=B547207CE5CE6F4EE3B52E7 0FB8C.
4. "Guideline for Intrapartum Care in Third Stage of Labour," National Institute for Health and Care Excellence (NICE), August 2021, nice.org.uk/guidance.
5. D. Farrar et al., "Care During the Third Stage of Labour: A Postal Survey of UK Midwives and Obstetricians," *BMC Pregnancy and Childbirth* 10, no. 23 (2010).
6. Sage-Femme Collective, "Natural Liberty: Rediscovering Self-Induced Abortion Methods," 2008, we.riseup.net/assets/351138/22321349-Natural-Liberty-Rediscovering -Self-Induced-Abortion-Methods.pdf.
7. R. T. Gunther, *The Greek Herbal of Dioscorides* (London: Hafner 1968), quoted in C. E. den Hertog, A. N. de Groot, and P. W. van Dongen, "History and Use of Oxy-tocics," *European Journal of Obstetrics & Gynecology and Reproductive Biology* 94, no. 1 (2001): 8–12.
8. S. Handley, "Abortion in the 19th Century," National Museum of Civil War Medi-cine, 2016, civilwarmed.org/abortion1/.
9. L. Schiebinger, "Exotic Abortifacients and Lost Knowledge," *The Lancet* 371 (March 1, 2008): 718–19.
10. E. West, "Reproduction and Resistance," in *Hidden Voices: Enslaved Women in the Lowcountry and U.S. South*, Lowcountry Digital History Initiative, ldhi.library .cofc.edu/exhibits/show/hidden-voices/resisting-enslavement/reproduction-and -resistance.

11. T. Haarmann et al., "Ergot: From Witchcraft to Biotechnology," *Molecular Plant Pathology* 10, no. 4 (2009): 563–77.

12. A. Lonitzer, *Kreuterbuch* (Frankfurt: Egenolff, 1482). Available online at digitale -sammlungen.de/de/view/bsb11200293?page=589.

13. Joachim Camerarius the Younger, *Commentary on Herbal Book of P. A. Mattioli* (1586). Available online at bildsuche.digitale-sammlungen.de/index.html?c=viewer &bandnummer=bsb00091089&pimage=00238&v=100&nav=.

14. Unknown author of *Codices Palatini* (Nuremberg, 1474). Available online at digi.ub .uni-heidelberg.de/diglit/cpg545/0144.

15. F. Rozier et al., *Journal de Physique, de chimie, d'histoire naturelle et des arts*, 1774. archive.org/details/journaldephysiq03unkngoog/page/144/mode/2up.

16. J.-B. Desgranges, "Sur la propriété qu'a le Seigle ergoté d'accélérer la marche de l'accouchement, et de hâter sa terminaison," *Nouveau Journal de Médecine* (1818), archive.org/details/BIUSante_90147x1818x01/page/n53/mode/2up.

17. J. Stearns, "Account of the Pulvis Parturiens, a Remedy for Quickening Child-birth," *Medical Repository* 2, no. 5 (January 1, 1808): 308–9. Available online at babel .hathitrust.org/cgi/pt?id=nyp.33433011578865&view=1up&seq=324&skin=2021.

18. Newsroom staff, "Medical Mysteries of Scotland's Medieval Hospital Unearthed," *The Scotsman*, October 25, 2017 (updated December 12, 2017).

19. R. Marya and R. Patel, *Inflamed: Deep Medicine and the Anatomy of Injustice* (London: Allen Lane, 2021), 188.

20. Stearns, "Account of the Pulvis Parturiens."

21. M. J. O'Dowd, *The History of Medications for Women* (New York: Parthenon, 2001).

22. Wellcome Collection, wellcomecollection.org/works/ehuwzq2d/items.

23. H. W. Dudley and C. Moir, "The Substance Responsible for the Traditional Clinical Effect of Ergot," *British Medical Journal* (March 16, 1935): 520–523.

24. K. Hofmann, *Vincent du Vigneaud 1901–1978: A Biographical Memoir* (Washington, DC: National Academy of Sciences, 1987). Available online at nasonline.org /publications/biographical-memoirs/memoir-pdfs/du-vigneaud-vincent.pdf.

25. H. H. Dale, "On Some Physiological Actions of Ergot," *Journal of Physiology* 34, no. 3 (1906).

26. G. H. Bell, *On Parturition and Some Related Problems of Reproduction* (Glasgow: University of Glasgow, 1943). Available online at proquest.com/openview/207bd 85ab4cba13ca52be52720c149d1/1?pq-origsite=gscholar&cbl=2026366&diss=y.

27. A. McLellan, "Response of Non-Gravid Human Uterus to Posterior-Pituitary Extract: and Its Fractions Oxytocin and Vasopressin," *The Lancet* (1940): 919–22.

28. E. H. Bishop, "Elective Induction of Labor," *Obstetrics & Gynecology* 5 (1955): 519–27.

29. E. Friedman, "The Graphic Analysis of Labor," *American Journal of Obstetrics and Gynecology* 68, no. 6 (1954): 1568–75.

30. D. J. MacRae, "Monitoring the Fetal Heart During a Pitocin Drip," Royal Society of Medicine Film Unit, 196? (exact year undocumented), accessed via the Wellcome Collection.

31. M. Nucci, A. R. Nakano, and L. A. Teixeira, "Synthetic Oxytocin and Hastening Labor: Reflections on the Synthesis and Early Use of Oxytocin in Brazilian Obstetrics," *História, Ciências, Saúde-Manguinhos* 25, no. 4 (October–December 2018): 979–98.

32. Ibid.

33. R. Reed, *Reclaiming Childbirth as a Rite of Passage: Weaving Ancient Wisdom with Modern Knowledge* (Yandina, Australia: Word Witch, 2021), 56.

34. E. C. Newnham, L. V. McKellar, and J. I. Pincombe, "Paradox of the Institution: Findings from a Hospital Labour Ward Ethnography," *BMC Pregnancy and Childbirth* 17, no. 1 (January 3, 2017): 2.

35. P. Middleton et al., "Induction of Labour at or Beyond 37 Weeks' Gestation," *Cochrane Database of Systematic Reviews* 7 (2020): art. no. CD004945.

36. H. G. Dahlen et al., "Intrapartum Interventions and Outcomes for Women and Children Following Induction of Labour at Term in Uncomplicated Pregnancies: A 16-Year Population-Based Linked Data Study," *BMJ Open* 11 (2021): e047040.

37. J. Agg, *The Uterus Monologues*, January 12, 2021, uterusmonologues.com/2021/01/12/birth-after-loss/.

38. M. Tolofari and L. Shepherd, "Postpartum Haemorrhage and Synthetic Oxytocin Dilutions in Labour," *British Journal of Midwifery* 29, no. 100 (2021): 590–6.

39. "Childbearing for Women Born in Different Years, England and Wales," Office for National Statistics, ons.gov.uk/peoplepopulationandcommunity/birthsdeathsandmarriages/conceptionandfertilityrates/bulletins/childbearingforwomenbornindifferentyearsenglandandwales/2019#childlessness.

40. G. Livingston, "They're Waiting Longer, but U.S. Women Today More Likely to Have Children Than a Decade Ago," Pew Research Center, January 18, 2018, pewresearch.org/social-trends/2018/01/18/theyre-waiting-longer-but-u-s-women-today-more-likely-to-have-children-than-a-decade-ago/.

41. "Campaign Against Painful Hysteroscopy," Hysteroscopy Action, hysteroscopyaction.org.uk.

42. S. Siricilla, C. C. Iwueke, and J. L. Herington, "Drug Discovery Strategies for the Identification of Novel Regulators of Uterine Contractility," *Current Opinion in Physiology* 13 (February 2020): 71–86.

43. E. E. Bafor and S. Kupittavanant, "Medicinal Plants and Their Agents That Affect Uterine Contractility," *Current Opinion in Physiology* 13 (2020): 20–26.

44. Reed, *Reclaiming Childbirth as a Rite of Passage*, 34.

Loss

1. F. Kahlo, in J. Espinoza, "Frida Kahlo's Last Secret Finally Revealed," *Guardian*, August 12, 2007.

2. "What Causes a Miscarriage?" Tommy's, tommys.org/baby-loss-support/miscarriage-information-and-support/causes-miscarriage.

3. L. Riverius et al., eds., *The Practice of Physick* (London: Peter Cole, 1658).

4. B. Jones and A. Shennan, "Cervical Cerclage," in H. Critchley, P. Bennett, and S. Thornton, eds., *Preterm Birth* (London: RCOG Press, 2004).

5. "Cervical Incompetence," Tommy's, tommys.org/pregnancy-information/pregnancy-complications/cervical-incompetence.

6. L. D. Tanner et al., "Maternal Race/Ethnicity as a Risk Factor for Cervical Insufficiency," *European Journal of Obstetrics and Gynecology and Reproductive Biology* 221 (2018): 156–9.

7. Tommy's, "Cervical Incompetence."

8. C-STICH2 trial information, ISRCTN registry. isrctn.com/ISRCTN12981869?q
=&filters=conditionCategory:Pregnancy%20and%20Childbirth,recruitmentCountry:
United%20Kingdom&sort=&offset=1&totalResults=338&page=1&pageSize=10
&searchType=basic-search.

9. K. Morris, email to author, October 4, 2021.

Cesarean

1. M. Cameron, "The Caesarean Section: With Notes of a Successful Case," *British Medical Journal* (January 26, 1889): 180–183.

2. "Caesarean Section—A Brief History: Part 1," U.S. National Library of Medicine, April 27, 1998 (updated July 26, 2013), nlm.nih.gov/exhibition/cesarean/part1
.html#:~:text=Perhaps%20the%20first%20written%20record,unable%20to%20
deliver%20her%20baby.

3. R. Dyce, "Case of Cæsarean Section," *Edinburgh Medical Journal* 7, no. 10 (1862): 895.

4. M. Cameron, "Caesarean Section and Its Modifications: With an Additional List of Five Cases," *Glasgow Hospital Reports* (1901). Available online at wellcomecollection
.org/works/hh4sbm2x/items?canvas=3.

5. "Births by Caesarean Section," World Health Organization, apps.who.int/gho/data
/node.main.BIRTHSBYCAESAREAN?lang=en.

6. "WHO Statement on Caesarean Section Rates," World Health Organization, 2015, WHO_RHR_15.02_eng.pdf;jsessionid=A673C403BE2860E7837A50BAB
A2DD855.

7. "NHS Maternity Statistics, England—2020–21," NHS Digital, digital.nhs.uk/data
-and-information/publications/statistical/nhs-maternity-statistics/2020-21.

8. J. Weaver and J. Magill-Cuerden, "'Too Posh to Push': The Rise and Rise of a Catch-phrase," *Birth* 40, no. 4 (2013): 264–71.

9. J. J. Weaver, H. Statham, and M. Richards, "Are There 'Unnecessary' Cesarean Sections? Perceptions of Women and Obstetricians About Cesarean Sections for Non-clinical Indications," *Birth* 34, no. 1 (March 2007): 32–41.

10. "Cesarean Delivery on Maternal Request," American College of Obstetricians and Gynecologists, January 2019, acog.org/clinical/clinical-guidance/committee
-opinion/articles/2019/01/cesarean-delivery-on-maternal-request.

11. "NICE Guideline 192: Caesarean Birth," National Institute for Health and Care Excellence (NICE), March 31, 2021, nice.org.uk/guidance/ng192/chapter
/Recommendations#maternal-request-for-caesarean-birth.

12. M. Jolly and J. Dunkley-Bent, "Letter on Use of Caesarean Section Rates Data," February 15, 2022.

13. R. Negrini et al., "Reducing Caesarean Rates in a Public Maternity Hospital by Implementing a Plan of Action: A Quality Improvement Report," *BMJ Open Quality* 9 (2020): e000791.

14. M. Lopes, "Caesarean Sections in Brazil Are an Audience Spectacle, with Wedding-Style Parties," *Washington Post*, June 12, 2019.

15. J. E. Potter et al., "Unwanted Caesarean Sections Among Public and Private Patients in Brazil: Prospective Study," *British Medical Journal* 323, no. 7322 (2001): 1155–8.

16. O. Khazan, "Why Most Brazilian Women Get C-Sections," *The Atlantic*, April 14, 2014.

17. S. Vedam et al., "The Giving Voice to Mothers Study: Inequity and Mistreatment During Pregnancy and Childbirth in the United States," *Reproductive Health* 16 (2019): 77.

18. R. Perez D'Gregorio, "Obstetric Violence: A New Legal Term Introduced in Venezuela," *International Journal of Gynecology and Obstetrics* 111, no. 3 (December 2010): 201–2.

19. G. Sen, B. Reddy, and A. Iyer, "Beyond Measurement: The Drivers of Disrespect and Abuse in Obstetric Care," *Reproductive Health Matters* 26, no. 53 (2018): 6–18.

20. V. Perrotte, A. Chaudhary, and A. Goodman, "'At Least Your Baby Is Healthy' Obstetric Violence or Disrespect and Abuse in Childbirth Occurrence Worldwide: A Literature Review," *Open Journal of Obstetrics and Gynecology* 10 (2020): 1544–62.

21. J. Smith, F. Plaat, and N. M. Fisk, "The Natural Caesarean: A Woman-Centred Technique," *British Journal of Obstetrics and Gynaecology* 115, no. 8 (2008): 1037–42.

22. S. Posthuma et al., "Risk and Benefits of a Natural Caesarean Section—a Retrospective Cohort Study," *American Journal of Obstetrics and Gynecology*, supplement to January 2015, S346.

23. N. Zafran et al., "The Impact of 'Natural' Cesarean Delivery on Peripartum Maternal Blood Loss: A Randomized Controlled Trial," *American Journal of Obstetrics and Gynecology*, supplement to January 2019, S630.

24. K. Bronsgeest et al., "Short Report: Post-Operative Wound Infections After the Gentle Caesarean Section," *European Journal of Obstetrics & Gynecology and Reproductive Biology* 241 (2019): 131–32.

25. S. Young, "Women Who Have 'Natural' C-Section Bond More with Their Baby, Say Doctors," *The Independent*, June 5, 2017.

26. R. Armbrust et al., "The Charité Cesarean Birth: A Family Orientated Approach of Cesarean Section," *Journal of Maternal-Fetal & Neonatal Medicine* 29, no. 1 (2016): 163–68.

27. R. Webb, S. Ayers, and A. Bogaerts, "When Birth Is Not as Expected: A Systematic Review of the Impact of a Mismatch Between Expectations and Experiences," *BMC Pregnancy and Childbirth* 21, no. 475 (2021).

28. V. Tonei, "Mother's Mental Health After Childbirth: Does the Delivery Method Matter?" *Journal of Health Economics* 63 (2019): 182–96.

29. E. Evans and M. Kupper, "Humanising Obstetric Care in Operating Theatres," *thebmjopinion* blog, *British Medical Journal*, April 22, 2021, blogs.bmj.com/bmj/2021/04/22/humanising-obstetric-care-in-operating-theatres/.

30. N. Fisk, F. Plaat, and J. Smith, "Natural Caesarean—a Decade On," Positive Birth Movement, July 30, 2018, positivebirthmovement.org/natural-caesarean-a-decade-on/.

31. Ibid.

32. R. Yoder, *Nightbitch* (London: Harvill Secker, 2021), 237.

Postpartum

1. A. Athan, "Matrescence," matrescence.com.

2. T. Mercado, "La Matriz Birth Services," lamatrizbirth.com/postpartum-sealing.

3. C. L. Dennis et al., "Traditional Postpartum Practices and Rituals: A Qualitative Systematic Review," *Women's Health* 3, no. 4 (July 2007): 487–502.

4. K. Mahabir, "Traditional Health Beliefs and Practices of Postnatal Women in Trinidad," dissertation for the University of Florida, 1997. Available online at ufdc.ufl.edu/AA00048623/00001/163j.

5. B. Layla, "Closing the Bones (Al Shedd), the Moroccan Way!" June 26, 2018, http:laylab.co.uk/tnp-blog/moroccanclosingthebones.

6. Fraser and Cooper, *Myles Texbook for Midwives*, 656.

7. S. Nashar et al., "Puerperal Uterine Involution According to the Method of Delivery," *Akush Ginekol* 46, no. 9 (2007): 14–18. Bulgarian. PMID: 18642558.

8. H. Negishi et al., "Changes in Uterine Size After Vaginal Delivery and Cesarean Section Determined by Vaginal Sonography in the Puerperium," *Archives of Gynecology and Obstetrics* 263, no. 1–2 (November 1999): 13–16.

9. "Core Restore Postpartum Belly Band," Lola & Lykke, lolalykke.com/products/core-restore-postpartum-support-band.

10. "Post-Pregnancy Belly Band," MammaBump, mammabump.com.

11. "Brenda S" on Amazon, January 23, 2018, amazon.com/ChongErfei-Postpartum-Support-Recovery-Shapewear.

12. "Post-Pregnancy Belly Band," MammaBump.

13. I. Karaca et al., "Influence of Abdominal Binder Usage After Cesarean Delivery on Postoperative Mobilization, Pain and Distress: A Randomized Controlled Trial," *Eurasian Journal of Medicine* 51, no. 3 (2019): 214–18.

14. S. Ghana et al., "Randomized Controlled Trial of Abdominal Binders for Postoperative Pain, Distress, and Blood Loss After Cesarean Delivery," *International Journal of Gynecology and Obstetrics* 137, no. 3 (June 2017): 271–76.

15. J. M. Szkwara et al., "Effectiveness, Feasibility, and Acceptability of Dynamic Elastomeric Fabric Orthoses (DEFO) for Managing Pain, Functional Capacity, and Quality of Life During Prenatal and Postnatal Care: A Systematic Review," *International Journal of Environmental Research and Public Health* 16, no. 13 (July 6, 2019): 2408.

16. G. Donnelly, email to author, January 6, 2022.

17. B. Davies, email to author, January 7, 2022.

18. J. Thomé, "I Tried Postpartum Belly Binding and Here's What Happened," Mom.com, May 30, 2019, mom.com/baby/202232-i-tried-postpartum-belly-binding-and-heres-what-happened.

Health

1. "Uterine Cancer Statistics," Cancer Research UK, cancerresearch.uk.org/health-professional/cancer-statistics/statistics-by-cancer-type/uterine-cancer#.

2. "Uterine Cancer: Statistics," Cancer.Net, cancer.net/cancer-type/uterine-cancer/statistics.

3. "Uterine Cancer Statistics," Cancer Research UK.

4. Cervical Cancer Action for Elimination, 2021, cervicalcanceraction.org.

5. "Cervical Cancer," Global Surgery Foundation, 2022, http:globalsurgeryfoundation.org/cervical-cancer.

6. "Guidelines for the Prevention and Early Detection of Cervical Cancer," American Cancer Society, April 22, 2012, cancer.org/cancer/cervical-cancer/detection-diagnosis-staging/cervical-cancer-screening-guidelines.

7. "When You'll Be Invited for Cervical Screening," NHS, nhs.uk/conditions/cervical-screening-when-youll-be-invited.

8. N. Chantziantoniou, "Lady Andromache (Mary) Papanicolaou: The Soul of Gynecological Cytopathology," *Journal of the American Society of Cytopathology* 3, no. 6 (2014): 319–26.

9. E. Kiourktsi, "Lifesaver," *Greece Is* (December 25, 2017): 104–7.

10. G. N. Papanicolaou and H. F. Traut, "The Diagnostic Value of Vaginal Smears in Carcinoma of the Uterus," *American Journal of Obstetrics and Gynecology* 42, no. 2 (1941): 193–206.

11. "Cervical Cancer Screening (PDQ—Health Professional Version)," National Cancer Institute, August 25, 2021, cancer.gov/types/cervical/hp/cervical-screening-pdq.

12. I. Pinnell, "Behind the Headlines: HPV Self-Sampling," Jo's Cervical Cancer Trust, February 24, 2021, jostrust.org/uk/about-us/news-and-blog/blog/behind-headlines -hpv-self-sampling.

13. "HPV Vaccination," Cervical Cancer Action, cervicalcanceraction.org.

14. "Cervical Cancer Elimination Initiative," World Health Organization, who.int /initiatives/cervical-cancer-elimination-initiative.

15. "YouScreen: Cervical Screening Made Easier," Small C, 2022, smallc.org.uk/get -involved-youscreen.

16. "Three Quarters of Sexual Violence Survivors Feel Unable to Go for Potentially Life-Saving Test," Jo's Cervical Cancer Trust, August 31, 2018, jostrust.org.uk/node/1075195.

17. "The Impact of Trauma and Cervical Screening," Somerset and Avon Rape and Sexual Abuse Support, June 14, 2021, sarsas.org.uk/cervical-screening.

18. A. M. Berner et al., "Attitudes of Transgender Men and Non-Binary People to Cervical Screening: A Cross-Sectional Mixed-Methods Study in the UK," *British Journal of General Practice* 71, no. 709 (2021): e614–e625.

19. "Screening and Treatment of Precancerous Lesions," Cervical Cancer Action, cervical canceraction.org/screening-and-treating-precancerous-lesions.

20. Global Surgery Foundatio, "Cervical Cancer."

21. "Supporting Our Sisters: Transforming Uterine Fibroid Awareness into Action," Society for Women's Health Research, March 23, 2021, swhr.org/event/supporting -our-sisters-transforming-uterine-fibroid-awareness-into-action/.

22. M. S. Ghant et al., "Beyond the Physical: A Qualitative Assessment of the Burden of Symptomatic Uterine Fibroids on Women's Emotional and Psychosocial Health," *Journal of Psychosomatic Research* 78, no. 5 (May 2015): 499–503.

23. S. E. Chiuve et al., "Uterine Fibroids and Incidence of Depression, Anxiety and Self-Directed Violence: A Cohort Study," *Journal of Epidemiology and Community Health* 76, no. 1 (2022): 92–99.

24. G. Roberts-Grey, "The Feelings Behind Our Fibroids," *Essence*, October 27, 2020.

25. R. Boynton-Jarrett et al., "Abuse in Childhood and Risk of Uterine Leiomyoma: The Role of Emotional Support in Biologic Resilience," *Epidemiology* 22, no. 1 (January 2011): 6–14.

26. H. Hutcherson, "Black Women Are Hit Hardest by Fibroid Tumors," *New York Times*, April 15, 2020.

27. D. D. Baird et al., "High Cumulative Incidence of Uterine Leiomyoma in Black and White Women: Ultrasound Evidence," *American Journal of Obstetrics and Gynecology* 188, no. 1 (2003): 100–107.

28. R. Myles, "Unbearable Fruit: Black Women's Experiences with Uterine Fibroids," dissertation for Georgia State University, 2013. Available online at scholarworks .gsu.edu/cgi/viewcontent.cgi?article=1071&context=sociology_diss.

29. S. T. Jones, "Uterine Fibroids: A Silent Epidemic," *The Hill*, June 6, 2007, thehill. com/homenews/news/12121-uterine-fibroidsa-silent-epidemic.

30. L. Dunham, "In Her Own Words: Lena Dunham on Her Decision to Have a Hysterectomy at 31," *Vogue*, February 14, 2018.

31. "Endometriosis Facts and Figures," Endometriosis UK, endometriosis-uk.org /endometriosis-facts-and-figures#1.

32. W. W. Russell, "Johns Hopkins Hospital Bulletin," vol. 10, pp. 8–10, quoted in G. Hannant, "Endometriosis: 1881–1940: The Discovery, Naming, Framing and Understanding of a Complicated Condition," B.Sc. dissertation for the University of London, 2002. Available online at wellcomecollection.org/works/etvep4bg.

33. J. A. Sampson, "Metastatic or Embolic Endometriosis, Due to the Menstrual Dissemination of Endometrial Tissue into the Venous Circulation," *American Journal of Pathology* 3, no. 2 (1927): 93–110.

34. Quoted in Hannant, "Endometriosis: 1881–1940," 523.

35. D. Redwine, "Mulleriosis Not Mullerianosis," letter commenting on P. G. Signorile et al., "Ectopic Endometrium in Human Foetuses Is a Common Event and Sustains the Theory of Müllerianosis in the Pathogenesis of Endometriosis, a Disease That Predisposes to Cancer," *Journal of Experimental & Clinical Cancer Research* (May 13, 2009), jeccr.biomedcentral.com/articles/10.1186/1756-9966-28-49/comments.

36. P. G. Signorile et al., "Ectopic Endometrium in Human Foetuses Is a Common Event and Sustains the Theory of Müllerianosis in the Pathogenesis of Endometriosis, a Disease That Predisposes to Cancer," *Journal of Experimental & Clinical Cancer Research* 28, no. 1 (April 9, 2009): 49.

37. D. O. Meike Schuster and D. A. Mackeen, "Fetal Endometriosis: A Case Report," *Fertility and Sterility* 103, no. 1 (January 2015): 160–162.

38. L. Osborne-Crowley, "A Common Treatment for Endometriosis Could Actually Be Making Things Worse," *Guardian*, July 2, 2021.

39. O. Bougie et al., "Influence of Race/Ethnicity on Prevalence and Presentation of Endometriosis: A Systematic Review and Meta-Analysis," *British Journal of Obstetrics and Gynaecology* 126, no. 9 (August 2019): 1104–15.

40. L. V. Farland and A. W. Horne, "Disparity in Endometriosis Diagnoses Between Racial/Ethnic Groups," *British Journal of Obstetrics and Gynaecology* (May 21, 2019): 1115–16.

41. A. Norman, *Ask Me About My Uterus* (New York: Bold Type Books, 2018), 19.

42. "BBC Research Announced Today Is a Wake-Up Call to Provide Better Care for the 1.5 Million with Endometriosis," Endometriosis UK, October 7, 2019, endometriosis-uk.org/news/bbc-research-announced-today-wake-call-provide -better-care-15-million-endometriosis-37606.

43. L. Hazard, *What the Midwife Said* podcast, season 1, episode 4, November 24, 2020, open.spotify.com/episode/2zUEA0NusEx0bDTQAGgnjJ?si=fgszmfuzRF22x NsyKwolqw.

44. K. Young, J. Fisher, and M. Kirkman, "Do Mad People Get Endo or Does Endo Make You Mad? Clinicians' Discursive Constructions of Medicine and Women with Endometriosis," *Feminism & Psychology* 29, no. 3 (2019): 337–56.

45. Clip from *Don Lemon Tonight*, August 8, 2015, CNN, edition.cnn.com/videos /us/2015/08/08/donald-trump-megyn-kelly-blood-lemon-intv-ctn.cnn.

46. H. D. Betz, *The Greek Magical Papyri in Translation, Including the Demotic Spells* (Chicago: University of Chicago Press, 1992), quoted in K. Marino, *Setting the Womb*

in Its Place: Toward a Contextual Archaeology of Graeco-Egyptian Uterine Amulets, doctoral dissertation for Brown University, March 2010. Available online at https://repository.library.brown.edu/studio/item/bdr:11094/PDF/.

47. E. Wright, "Magic to Heal the 'Wandering Womb' in Antiquity," Folklore Thursday, January 18, 2018, folklorethursday.com/folklife/magic-to-heal-the-wandering-womb-in-antiquity/.

48. L. Rivière, *The secrets of the famous Lazarus Riverius, councellor & physician to the French king, and professor of physick in the University of Montpelier newly translated from the Latin by E.P., M.D.* Available online from the Text Creation Partnership at name.umdl.umich.edu/A57364.0001.001.Prat, E. p. 73.

49. C. Tasca et al., "Women and Hysteria in the History of Mental Health," *Clinical Practice and Epidemiology in Mental Health* 8 (2012): 110–19.

50. A. Hustvedt, *Medical Muses* (London: Bloomsbury, 2011).

51. *The Ladies Dispensatory* (London: Printed for James Hodges and John James, 1739). Available online at wellcomecollection.org/works/m3kfwmyk.

52. J. Strohecker, "A New Vision of Wellness," Healthy.net, September 24, 2019, healthy.net/2019/09/24/a-new-vision-of-wellness/.

53. J. Parvati, *Hygieia: A Woman's Herbal* (Berkeley, CA: Freestone, 1978), 99.

54. Ibid., ix.

55. Ibid., glossary.

56. S. Callaghan et al., "The Future of the $1.5 Trillion Wellness Market," McKinsey, April 8, 2021, mckinsey.com/industries/consumer-packaged-goods/our-insights/feeling-good-the-future-of-the-1-5-trillion-wellness-market.

57. Alice (last name withheld at interviewee's request), online message to author, June 1, 2021.

58. "The Infrared Sauna and Detox Spa Guide," Goop, goop.com/city-guide/infrared-saunas-detox-spas-and-the-best-spots-for-colonics/tikkun-spa.

59. Ibid.

60. "Pu$$y Power Rose Quartz Infused Yoni & Vaginal Wash," Goddess Detox, goddessdetox.org/collections/self-love-inspired-products/products/pu-y-power-crystal-infused-yoni-vaginal-wash?variant=39370179084336.

61. Femmagic, femmagic.com.

62. "Queen Tings Yoni & Vagina Steaming Gown," Goddess Detox, goddessdetox.org/collections/self-love-inspired-products/products/queen-tings-yoni-vaginal-steaming-gown?variant=32337904042032.

63. "Yoni Steam Herbs: Women's Blend," The Plant Path Folk, theplantpathfolk.co.uk/apothecary.

64. A. Trivedi, reply to author's post on Twitter, June 15, 2021.

65. J. Gunter, "No GOOP, We Are Most Definitely Not on the Same Side," personal blog, July 26, 2019, drjengunter.com/2019/07/26/no-goop-we-are-most-definitely-not-on-the-same-side/.

66. C. Shea, "Jen Gunter on Why Vulvas Don't Need a Summer Glow-Up," Refinery29, June 2, 2021, refinery29.com/en-ca/2021/06/10445943/jen-gunter-menopause-manifesto-vagina-glow-up.

67. Dr. Jen Gunter on "Vagina Profiteers: The Economics of the Wellness Industrial Complex," Gender and the Economy, gendereconomy.org/dr-jen-gunter/.

68. N. Ding, S. Batterman, and S. K. Park, "Exposure to Volatile Organic Compounds and Use of Feminine Hygiene Products Among Reproductive-Aged Women in the United States," *Journal of Women's Health* 29, no. 1 (2020): 65–73.

69. J. Zhang, A. G. Thomas, and E. Leybovich, "Vaginal Douching and Adverse Health Effects: A Meta-Analysis," *American Journal of Public Health* 87 (1997): 1207–11.

70. T. Vandenburg and V. Braun, "Basically, It's Sorcery for Your Vagina: Unpacking Western Representations of Vaginal Steaming," *Culture, Health & Sexuality* 19, no. 4 (October 10, 2016): 472.

71. Ibid., 480.

72. M. Fricker, introduction to *Epistemic Injustice: Power and the Ethics of Knowing* (Oxford: Oxford University Press, 2007). Available online at mirandafricker.com/uploads /1/3/6/2/136236203/introduction.pdf.

73. A. Lorde, *A Burst of Light* (Ann Arbor, MI: Firebrand Books, 1988).

Menopause

1. Y. M. Baron, *A History of the Menopause*, University of Malta, 2012. Available online at researchgate.net/publication/304346490_A_History_of_the_Menopause.

2. C. P. L. De Gardanne, *Avis aux femmes qui entrent dans l'age critique* (Paris: Imprimerie de J. Moronval, 1816). Available online at wellcomecollection.org/works /utrvvj2v/items?canvas=9.

3. C. P. L. De Gardanne, *De la menopause: Ou de l'age critique des femmes* (1821).

4. A. Strachey, ed., *The Standard Edition of the Complete Psychological Works of Sigmund Freud, Vol. 12 (1911–1913)* (London: Vintage, 1958), quoted in P. Maddison, "Reclaiming Menopause from the Medics," *Contemporary Psychotherapy* 11, no. 2 (2019). Available online at contemporarypsychotherapy.org/volume-11-issue-2-winter-2019 /reclaiming-menopause-from-the-medics/.

5. H. Deutsch, *The Psychology of Women* (New York: Grune and Stratton, 1958), quoted in T. M. Luhrmann, "Review of *The Slow Moon Climbs* by Susan P. Mattern," *Times Literary Supplement*, March 13, 2020.

6. R. Wilson, *Feminine Forever* (New York: M. Evans, 1968), quoted in T. M. Luhrmann, "Review of *The Slow Moon Climbs*."

7. M. Doughty, "Case Study: The Medical Menopause," *Bodies of Difference*, December 26, 2016, thedifferenceofbodies.wordpress.com/2016/12/06/75/.

8. T. Eytan, "Pharmaceutical Ads from the 20th Century," Flickr, January 14, 2018, flickr.com/photos/taedc/38798081665.

9. R. Benaroch, "Premarin—How Marketing Popularized Treatment for Menopausal Symptoms," Wondrium Daily, April 29, 2019, wondriumdaily.com/premarin-how -marketing-popularized-treatment-for-menopausal-symptoms.

10. P. Waller-Bridge, *Fleabag*, season 2, episode 3, first aired on the BBC on March 18, 2019.

11. U. K. Le Guin, "The Space Crone," in R. Formanek, ed., *The Meanings of Menopause* (London: Routledge, 1990), xxiii.

Hysterectomy

1. J. D. Wright et al., "Nationwide Trends in the Performance of Inpatient Hysterectomy in the United States," *Obstetrics & Gynecology* 122, no. 2, part 1 (2013): 233–41.

2. T. Cornforth, "Facts About Hysterectomy in the United States," Verywell Health, November 25, 2020, verywellhealth.com/the-facts-about-hysterectomy-in-the-united -states-3520837.

3. P. Willughby, *Observations in Midwifery: as also The Country Midwifes Opusculum or Vade Mecum* (Warwick: H. T. Cooke and Son, 1863), 251–2. Available online at archive.org/details/observationsinmi00will/page/n5/mode/2up.

4. T. Keith, *Contributions to the Surgical Treatment of Tumours of the Abdomen*, vol. 1, quoted in C. Sutton, "Hysterectomy: A Historical Perspective," *Baillière's Clinical Obstetrics and Gynaecology* 11, no. 1 (March 1997): 1–22.

5. Y. Savage, email to author, April 30, 2021.

6. D. Whelan, email to author, April 29, 2021.

7. "Stephanie" (identifying details changed at interviewee's request), online message to author, April 28, 2021.

8. "Natalya" (identifying details changed at interviewee's request), email to author, April 29, 2021.

9. J. Forst, "Study Finds Women at Greater Risk of Depression, Anxiety After Hysterectomy," Mayo Clinic News Network, September 4, 2019, newsnetwork.mayoclinic .org/discussion/study-finds-women-at-greater-risk-of-depression-anxiety-after -hysterectomy/.

10. "Hysterectomy—Recovery," NHS. nhs.uk/conditions/hysterectomy/recovery/.

11. The University of Arizona, arizona.edu/about.

12. S. V. Koebele et al., "Hysterectomy Uniquely Impacts Spatial Memory in a Rat Model: A Role for the Nonpregnant Uterus in Cognitive Processes," *Endocrinology* 160, no. 1 (January 1, 2019): 1–19.

13. G. Enders, *Gut: The Inside Story of Our Body's Most Underrated Organ (Revised Edition)* (London: Scribe, 2017).

14. L. E. Corona et al., "Use of Other Treatments Before Hysterectomy for Benign Conditions in a Statewide Hospital Collaborative," *American Journal of Obstetrics and Gynecology* 212, no. 3 (March 2015): 304.e1–7.

15. "Nearly One in Five Women Who Undergo Hysterectomy May Not Need the Procedure," Elsevier, January 6, 2015, elsevier.com/about/press-releases/archive /research-and-journals/nearly-one-in-five-women-who-undergo-hysterectomy -may-not-need-the-procedure.

16. World Professional Association for Transgender Health, "Standards of Care for the Health of Transsexual, Transgender, and Gender Nonconforming People," 7th version, 2012, wpath.org/publications/soc.

17. I. T. Nolan, C. J. Kuhner, and G. W. Dy, "Demographic and Temporal Trends in Transgender Identities and Gender Confirming Surgery," *Translational Andrology and Urology* 8, no. 3 (2019): 184–90.

18. S. E. James et al., *The Report of the 2015 U.S. Transgender Survey* (Washington, DC: National Center for Transgender Equality, 2016).

19. *Oprah Winfrey Show*, April 15, 2008.

20. V. Parsons, "Academic Says Pregnancy Is 'Masculine' as It's Revealed 22 Transgender Men Gave Birth in Australia Last Year," Pink News, August 15, 2019, pinknews.co.uk/2019/08/15/22-transgender-men-gave-birth-in-australia-last -year-pregnancy/.

21. R. Pearce, *Understanding Trans Health* (Bristol: Policy Press, 2018), 27.
22. Women and Equalities Committee, House of Commons, *Oral Evidence: Reform of the Gender Recognition Act HC129*, May 12, 2021, committees.parliament.uk/oral evidence/2177/html/.

Reprocide

1. C. Lamb, *Our Bodies, Their Battlefield* (London: William Collins, 2021).
2. S. Zhang, "J. Marion Sims: the Gynecologist Who Experimented on Slaves," *The Atlantic*, April 18, 2018.
3. J. M. Sims (H. Marion-Sims, ed.), *The Story of My Life* (New York: D. Appleton 1884). Available online at babel.hathitrust.org/cgi/pt?id=hvd.32044013687306&v iew=1up&seq=9&skin=2021.
4. D. Ojanuga, "The Medical Ethics of 'the Father of Gynaecology,' Dr. J. Marion Sims," *Journal of Medical Ethics* 19 (1993): 28–31.
5. F. Galton, *Hereditary Genius: An Inquiry into Its Laws and Consequences* (London: Macmillan, 1869).
6. N. Antonios and C. Raup, "Buck v. Bell (1927)," Embryo Project Encyclopedia, January 1, 2012, embryo.asu.edu/pages/buck-v-bell-1927.
7. A. Cohen, *Imbeciles: The Supreme Court, American Eugenics, and the Sterilization of Carrie Buck* (New York: Penguin Press, 2016), 24.
8. US Supreme Court, "BUCK v. BELL, Superintendent of State Colony Epileptics and Feeble Minded," 1927. Available online at law.cornell.edu/supremecourt /text/274/200.
9. R. J. Cynkar, "Buck v. Bell: 'Felt Necessities' v. Fundamental Values?" *Columbia Law Review* 81, no. 7 (1981): 1418–61.
10. US Supreme Court, "BUCK v. BELL," 1927.
11. P. Lombardo, transcript of "The Rape of Carrie Buck," Cold Spring Harbor Laboratory DNA Learning Center, 2020, dnalc.cshl.edu/view/15234/The-rape-of-Carrie -Buck-Paul-Lombardo.html.
12. E. Black, "Eugenics and the Nazis—the California Connection," *SFGate*, September 9, 2003, sfgate.com/opinion/article/Eugenics-and-the-Nazis-the-California -2549771.php.
13. A. Hitler, *Mein Kampf*, 1924, quoted in ibid.
14. Ibid.
15. C. Clauberg, letter to H. Himmler, June 7, 1943, from "Nazi Letters on Sterilization," Remember.org, remember.org/witness/links-let-ster.
16. Ibid.
17. S. Benedict and J. Georges, "Nurses and the Sterilization Experiments of Auschwitz: A Postmodernist Perspective," *Nursing Inquiry* 13, no. 4 (December 2006): 277–88.
18. US Supreme Court, "SKINNER v. STATE OF OKLAHOMA ex rel. WILLIAMSON, Atty. Gen. of Oklahoma," 1942. Available online at law.cornell.edu/supreme court/text/316/535.
19. S. Sebring, "Sterilization—Japanese American Women," *Mississippi Appendectomy*, November 25, 2007, mississippiappendectomy.wordpress.com/2007/11/25 /sterilization-japanese-american-women/.

20. S. Garcia, "8 Shocking Facts About Sterilization in U.S. History," *Mic*, October 7, 2013, mic.com/articles/53723/8-shocking-facts-about-sterilization-in-u-s-history.

21. "Fannie Lou Hamer," PBS, *American Experience*, pbs.org/wgbh/americanexperience /features/freedomsummer-hamer/.

22. L. J. Ross and R. Solinger, *Reproductive Justice* (Berkeley: University of California Press, 2017), 50–51.

23. T. B. Brown, "Who Are the Confederate Men Memorialized with Statues?" NPR, August 18, 2017, npr.org/2017/08/18/543626600/who-are-the-confederate-men -memorialized-with-statues?t=1638887435968.

24. J. Heim, "How Is Slavery Taught in America? Schools Struggle to Teach It Well," *Washington Post*, August 28, 2019.

25. "Jefferson Davis Memorial State Historic Site," Department of Natural Resources Division, Georgia State Parks, gastateparks.org/jeffersondavismemorial.

26. House of Representatives, 104th Congress, 2nd session, "Illegal Immigration Reform and Immigrant Responsibility Act of 1996," September 24, 1996, congress .gov/104/crpt/hrpt828/CRPT-104hrpt828.pdf.

27. Oldaker v. Giles, District Court, M. D. Georgia, August 4, 2021, casetext.com/case /oldaker-v-giles.

28. "Imprisoned Justice," Project South, June 2017, projectsouth.org/wp-content/uploads /2017/06/Imprisoned_Justice_Report-1.pdf.

29. "Complaint Re: Lack of Medical Care, Unsafe Work Practices, and Absence of Adequate Protection Against COVID-19 for Detained Immigrants and Employees Alike at the Irwin County Detention Center," Project South, September 14, 2020, projectsouth.org/wp-content/uploads/2020/09/OIG-ICDC-Complaint -1.pdf.

30. Ibid.

31. "We Stand with Mahendra Amin," Facebook page, facebook.com/We-Stand-With -Mahendra-Amin-109571914226828/.

32. N. Merchant, "Migrant Women to No Longer See Doctor Accused of Misconduct," Associated Press, September 22, 2020, apnews.com/article/georgia-archive -immigration-f3b1007a9d2ef3cb6d2bd410673eae83.

33. *Belly of the Beast*, ITVS and Idlewild Films, 2020.

34. E. Cohn, email to the author, May 20, 2021.

35. "Czech Republic: Hard Won Justice for Women Survivors of Unlawful Sterilization," Amnesty International, July 22, 2021, amnesty.org/en/latest/news /2021/07/czech-republic-hard-won-justice-for-women-survivors-of-unlawful -sterilization/.

36. "China Cuts Uighur Births with IUDs, Abortion, Sterilization," Associated Press, June 29, 2020, apnews.com/article/ap-top-news-international-news-weekend-reads -china-health-269b3de1af34e17c1941a514f78d764c.

37. L. Ross, "Conceptualizing Reproductive Theory: A Manifesto for Activism," in L. Ross et al., eds., *Radical Reproductive Justice: Foundation, Theory, Practice, Critique* (New York: Feminist Press, 2017), eBook location 3506.

38. Ibid.

39. Martin Luther King Jr.'s exact quote, from a speech given at the National Cathedral in Washington, DC, on March 31, 1968, was, "We shall overcome because the arc of

the moral universe is long but it bends toward justice." Available online at youtube
.com/watch?v=AFbt7cO30jQ.

40. A. Shahshahani, Twitter post on May 20, 2021, twitter.com/ashahshahani/status
/1395378848498339840.

Future

1. M. Brännström, L. Johannesson, H. Bokström, et al., "Livebirth After Uterus Transplantation," *Lancet* 385 (2015): 607–16.

2. M. Brännström et al., "Live Birth After Robotic-Assisted Live Donor Uterus Transplantation," *Acta Obstetrica Gynecologica Scandinavica* 99, no. 9 (September 2020):
1222–9.

3. M. Brännström, "The Swedish Uterus Transplantation Project: The Story Behind
the Swedish Uterus Transplantation Project," *Acta Obstetrica Gynecologica Scandinavica* 94 (2015): 675–9.

4. TEDx Talks, "The World's First Uterus Transplantation from Mother to Daughter:
Mats Brännström at TEDxGöteborg," December 27, 2013, youtube.com/watch
?v=60AJPw—qwk.

5. Ibid.

6. Ibid.

7. T. F. Murphy, "Assisted Gestation and Transgender Women," *Bioethics* 29, no. 6
(2015): 389–97.

8. S. Eraslan, R. J. Hamernik, and J. D. Hardy, "Replantation of Uterus and Ovaries in
Dogs, with Successful Pregnancy," *Archives of Surgery* 92, no. 1 (1966): 9–12.

9. M. Brännström, "Uterus Transplantation," *Current Opinion in Organ Transplantation*
20 (2015): 621–28.

10. Ibid.

11. H. Thomasy, "Scientists Think a Lab-Grown Uterus Could Help Fight Infertility,"
Future Human, February 3, 2021, futurehuman.medium.com/scientists-think-a
-lab-grown-uterus-could-help-fight-infertility-e263ab2e397d.

12. R. S. Magalhaes et al., "A Tissue-Engineered Uterus Supports Live Births in Rabbits," *Nature Biotechnology* 38 (2020): 1280–87.

13. E. Partridge et al., "An Extra-Uterine System to Physiologically Support the Extreme Premature Lamb," *Nature Communications* 8 (2017): 15112.

14. A. Huxley, *Brave New World* (London: Chatto & Windus, 1932).

15. K. Kingma and S. Finn, "Neonatal Incubator or Artificial Womb? Distinguishing
Ectogestation and Ectogenesis Using the Metaphysics of Pregnancy," *Bioethics* 34,
no. 4 (April 5, 2020): 354–63.

16. D. Begović et al., "Reviewing the Womb," *Journal of Medical Ethics* 47 (2021): 820–29.

17. S. Oelhafen et al., "Informal Coercion During Childbirth: Risk Factors and Prevalence Estimates from a Nationwide Survey of Women in Switzerland," *BMC Pregnancy and Childbirth* 21 (2021): 369.

18. N. Jeffay, "In Breakthrough, Israelis Grow Hundreds of Mouse Embryos in Artificial
Wombs," *Times of Israel*, March 17, 2021.

19. A. Aguilera-Castrejon et al., "Ex Utero Mouse Embryogenesis from Pre-Gastrulation
to Late Organogenesis," *Nature* 593 (2021): 119–24.

20. W. G. Kantor, "Woman Gave Birth Via Uterus Transplant," *People*, February 13, 2020.

21. J. Gobrecht, Instagram post, November 17, 2020.

22. C. Stewart, "Number of Deceased Organ Transplants in the UK 2020/21, by Organ Donated," Statista, July 28, 2021, statista.com/statistics/380145/number-of-organ-transplants-by-organ-donated-in-uk/.

23. "Transplant Safety," Centers for Disease Control and Prevention, cdc.gov/transplant safety/overview/key-facts.html.

24. A. Syrtash, *Pregnantish* podcast on Apple Podcasts, December 23, 2020, podcasts .apple.com/gb/podcast/meet-the-3rd-person-in-the-world-to-have-a/id1461336652 ?i=1000503354444.

25. A. L. Caplan et al., "Moving the Womb," *Hastings Center Report* 37, no. 3 (May–June 2007): 18–20.

26. Syrtash, *Pregnantish*.

27. S. Heti, *Motherhood* (London: Vintage, 2019), 44.

28. B. P. Jones et al., "Uterine Transplantation in Transgender Women," *British Journal of Obstetrics and Gynaecology* 126, no. 2 (2019): 152–56.

29. "The Harms of Denying a Woman a Wanted Abortion: Findings from the Turnaway Study," Advancing New Standards in Reproductive Health (ANSIRH), University of California, San Francisco (UCSF), April 16, 2020, ansirh.org/sites /default/files/publications/files/the_harms_of_denying_a_woman_a_wanted _abortion_4-16-2020.pdf.

30. "Abortion Factsheet," World Health Organization, 2022, who.int/health-topics /abortion#tab=tab_1.

31. S. Lale et al., "Global Causes of Maternal Death: A WHO Systematic Analysis," *Lancet Global Health* 2, no. 6 (2014): e323–e333.

32. "Abortion," World Health Organization, November 25, 2021, who.int/news-room /fact-sheets/detail/abortion.

33. Ibid.

34. "Investigation into the Safety, Quality and Standards of Services Provided by the Health Service Executive to Patients, Including Pregnant Women, at Risk of Clinical Deterioration, Including Those Provided in University Hospital Galway, and as Reflected in the Care and Treatment Provided to Savita Halappanavar," Health Information and Quality Authority, October 7, 2013, hiqa.ie/sites/default /files/2017-01/Patient-Safety-Investigation-UHG.pdf.

35. "Czestochowa. Agnieszka, 37, Died in the Hospital. Family: Decaying Bodies of Unborn Sons Were Left in It, the Hospital's Statement," *Polish News*, January 26, 2022, polishnews.co.uk/czestochowa-agnieszka-37-died-in-the-hospital-family-decaying -bodies-of-unborn-sons-were-left-in-it-the-hospitals-statement/.

36. Legislature of the State of Texas, S.B. No. 8, enacted September 1, 2021, capitol .texas.gov/tlodocs/87R/billtext/pdf/SB00008F.pdf.

37. Society for Maternal-Fetal Medicine, Advocacy Action Center, smfm.org/advocacy /vv?vsrc=%2fcampaigns%2f86901%2frespond.

38. Guttmacher Institute, "State Bans on Abortion Throughout Pregnancy," January 1, 2022, guttmacher.org/state-policy/explore/state-policies-later-abortions.

39. Guttmacher Institute, "Abortion Policy in the Absence of Roe," January 13, 2022, guttmacher.org/state-policy/explore/abortion-policy-absence-roe.

40. Alito, Samuel, for the Supreme Court of the United States, Dobbs v. Jackson Women's Health Organization, June 24, 2022, supremecourt.gov/opinions/21pdf/19 -1392_6j37.pdf.

41. Guttmacher Institute, "'Choose Life' License Plates," January 1, 2022, guttmacher .org/state-policy/explore/choose-life-license-plates.
42. M. Atwood, introduction to "The Network" by C. Pires, *Guardian* magazine, February 19, 2022, p. 29.
43. UNFPA, *My Body Is My Own: Claiming the Right to Autonomy and Self-Determination*, 2021, unfpa.org/SOWP-2021.
44. Ibid.

Abortifacient: (adj.) initiating or causing abortion, or (noun) a substance that has this effect.

Abortion: spontaneous or induced termination of a pregnancy; most commonly, the word is used to describe a termination achieved by pharmacological or surgical means.

Adenomyosis: a disease in which tissue similar to that of the endometrium (the lining of the womb) is found within the myometrium (the inner muscle layer of the womb).

Amenorrhea: the absence or cessation of menstruation. Amenorrhea can have a variety of causes, including but not limited to stress, overexertion, weight loss, illness, hormonal disturbance, or pregnancy.

Amnion: the inner membrane of the pregnancy sac or "bag of waters," containing the fetus and amniotic fluid.

Amniotic fluid: the fluid surrounding a fetus during pregnancy.

Antenatal: during pregnancy; before birth.

Antiseptic: preventing the survival and proliferation of potentially dangerous microorganisms such as bacteria and viruses.

Artery: a vessel carrying oxygenated blood away from the heart.

Blastocyst: a very early pregnancy structure, forming roughly five to seven days after conception. The outer layer of this ball of cells goes on to form the placenta and chorion, while the inner layer develops into the fetus and amnion.

Braxton Hicks contractions: sporadic, sometimes rhythmic tightenings of the uterus which are often felt in the latter stages of pregnancy. Generally painless but sometimes uncomfortable, Braxton Hicks contractions seldom initiate dilatation of the cervix.

Cervix: the "neck" or lowest part of the uterus; the thick, fleshy tube that connects the body of the uterus to the vagina. The cervix is able to tip forward and back within the pelvis, and can thin, soften, and dilate during birth.

Chorion: the outer membrane of the pregnancy sac or "bag of waters."

Cisgender: relating to or describing a person whose gender identity matches that which they were assigned at birth.

Clitoris: a sensitive, nerve-rich organ of erectile tissue located in the vulva. Previously, the clitoris was thought to be a small, hooded structure situated anterior to the opening of the vagina; it is now understood that this is simply the visible part of the clitoris, and that the organ itself extends inwardly along the labia in two *crurae* or bulblike roots.

Coitus: the act of sexual intercourse, usually referring to penetrative sex.

Contraction: a shortening of uterine muscle fibers, followed by a relaxation of these fibers. The uterus contracts imperceptibly throughout pregnancy, although in the latter stages of gestation, these tightenings can be felt as Braxton Hicks contractions, and in labor, these contractions aid first in effacing and dilating the cervix, and later in expelling the fetus, placenta, and membranes.

Corpus: the main body of the uterus.

Corpus luteum: following ovulation, the remaining part of the egg follicle left behind within the ovary. The corpus luteum produces estrogen and progesterone unless or until fertilization of a released egg fails to occur.

Discharge: normal physiological excretion which is expelled through the vagina as a mucusy substance. Discharge varies in color, consistency, amount, and odor throughout the menstrual cycle and throughout the female life cycle as a whole.

Doula: a layperson who provides nonclinical, emotional, and practical support to the birthing person at any time during the childbearing year, from pregnancy to birth and/or the postnatal weeks or months.

Dysmenorrhea: painful menstruation.

Ectogenesis: the gestation of a fetus outside a human uterus.

Ectopic pregnancy: a pregnancy which occurs outside the uterus (but still within the body); for example, in a uterine tube or elsewhere in the abdominal cavity. Ectopic pregnancies cannot safely be carried to term and can be life-threatening for the mother.

Embryo: a term to describe a mammal or human developing in the womb between roughly the second and eighth weeks of gestation.

Endometriosis: a disease in which tissue resembling that of the endometrium (the lining of the womb) adheres to structures outside the uterus, responding to fluctuating hormone levels and often causing pain, internal bleeding, and inflammation as it does so.

Endometrium: the lining of the uterus. During the menstrual cycle, the endometrium thickens in response to hormonal signals and is subsequently sloughed off and expelled.

Episiotomy: a deep cut made in the moments before birth to widen the opening of the vagina; routinely performed during forceps deliveries and also sometimes in unassisted vaginal births.

Epithelium: the layer of cells lining internal organs and vessels.

Ergometrine: an alkaloid derivative of ergot, a fungus found growing on rye plants. Ergometrine is used to prevent or manage postpartum hemorrhage, and can be administered orally, intravenously, or intramuscularly.

Estrogen: a hormone essential for female sexual development, menstrual function, the maintenance of pregnancy, and general health. Estrogen diminishes during menopause and is sometimes replaced or supplemented to aid well-being during this stage.

Fallopian tubes: *see* uterine tubes.

Fetus: the term used to describe a human infant developing within the uterus after eight weeks of gestation.

Fibroid: a benign fibrous mass or tumor that can be found growing in the uterine cavity or within the muscle layer of the uterus.

Fornix: an arch or "pocket" within the body; for example, the anterior vaginal fornix is a space or pocket that can be found at the top of the vagina, in front of the cervix, while the posterior vaginal fornix is a space or pocket that can be found at the top of the vagina, behind the cervix.

Fundus: the uppermost part of the uterus.

Gamete: a mature male or female reproductive cell which can join its counterpart from the opposite sex to form a fertilized pregnancy; a sperm or an egg.

Gynecology: the study of the female reproductive system; also, the branch of medicine involved with maintaining the health of this system and treating diseases relating to it.

Howdie: a Scots term for community-based lay midwives who often attended neighbors in their homes for both birth and death. Howdies were prolific until the Midwives Act (Scotland) 1915 introduced professional regulation.

Hysterectomy: surgical removal of the uterus.

Hysteria: a now-disproven theory that women's reproductive organs make them emotionally and mentally unstable.

Hysteroscopy: a procedure in which a thin telescopic device is passed through the cervix to allow visual inspection of the inside of the uterus.

IVF or in vitro fertilization: an assisted reproduction technique in which conception occurs in a dish or vessel (*in vitro*, or "in glass") in a laboratory using previously collected eggs and sperm.

Intrapartum: during labor and birth.

Ketones: compounds produced in the body when fat is burned for energy instead of glucose; for example, during starvation or excessive vomiting, or in some stages of uncontrolled diabetes.

Laparoscopy: a surgical procedure in which an internal camera is used to explore or visualize a part of the body, often alongside instruments which can be controlled remotely by the operator.

Labia: the fleshy folds or "lips" surrounding the vaginal opening. The thinner, innermost lips are the labia minora; the padded, outermost folds are the labia majora. There is a vast diversity in the appearance of normal labia, which can vary in color, thickness, length, and tone.

Ligament: a tough, fibrous band of connective tissue which joins two bones together or supports an internal organ such as the uterus.

Lower surgery: a gender confirming surgery in which a person's external genitalia are refashioned or reconstructed to match the sex with which they identify; for example, lower surgery for a trans man would involve refashioning of the vulva and vagina into a penis and scrotum. Conversely, a trans woman could opt for surgical creation of vulval and vaginal structures.

Luteal phase: the latter part of the menstrual cycle, following ovulation. In this phase, the sac around a released egg or ovum produces progesterone and estrogen, encouraging the lining of the womb to thicken in preparation for a potential pregnancy.

Mayer-Rokitansky-Küster-Hauser syndrome, or MRKH: a congenital condition in which the uterus, cervix, and sometimes vagina are underdeveloped or absent, caused by a variation in development of the Mullerian duct in early pregnancy.

Meconium: the first feces excreted by a fetus or newborn infant; a thick, tarry substance containing bile pigments, mucus, and epithelial cells from the lining of the gut.

Menarche: the onset of a person's first menstrual period.

Menopause: a term for the gynecological life stage reached twelve months after a person's final menstrual period.

Menorrhagia: heavy menstrual bleeding.

Menses: the time of a person's menstrual bleeding; also sometimes used to describe the flow itself.

Menstrual cycle: the hormonally controlled sequence of events in which the reproductively mature body of a person with biologically female anatomy prepares itself for a potential pregnancy each month. In this process, an egg is released from one of the ovaries, the lining of the womb thickens to prepare for potential fertilization and implantation, and if a pregnancy is not achieved, the egg and uterine lining are passed along with other components of menstrual effluent (see below) through the cervix and out of the vagina.

Menstrual effluent: the substance expelled via the vagina in menstruation, containing not only blood but also mucus, epithelial cells, microbes, inflammatory substances, and immune cells that can provide important information about the health of the menstruator.

Microbiome: a collective term for the microbes that exist intrinsically within an organ or physiological environment. A microbiome may include bacteria, viruses, fungi, yeasts, and many other microorganisms that contribute to both health and disease.

Midwifery: the art, science, and craft of caring for women and other birthing people throughout the childbearing year: prior to conception, during pregnancy, in labor, and in the first few weeks of early parenthood.

Miscarriage: early pregnancy loss, often referring to a pregnancy before the legal age of viability (broadly defined as twenty-four weeks of gestation, although advances in neonatal medicine have improved outcomes for some infants born even earlier).

Mullerian duct: an early embryonic structure that then recedes in the male, and differentiates to form a urogenital tract and external genitalia in the female.

Multipara or multip: a person who has given birth to more than one viable infant.

Myometrium: the inner muscle layer of the uterus.

Natural killer cells or NK cells: a kind of immune cell now known to be found at the interface of the endometrium and the placenta, a site of complex immunological balance.

Obstetrician/gynecologist or ob/gyn: a doctor who specializes in pregnancy, birth, and the female reproductive system.

Obstetrics: the medical study of pregnancy, birth, and the postnatal period.

Organoid: a lab-grown, three-dimensional structure that imitates the structure and function of tissue from a human organ such as a uterus or placenta.

Ovary: one of two small, almond-shaped organs located to either side of the uterus. Ovaries contain egg follicles which, under hormonal influence, mature into eggs capable of being fertilized by sperm.

Ovum or egg: the female reproductive cell, containing genetic material from the mother. In a normal menstrual cycle, an ovum is released each month during ovulation.

Oxytocic: a synthetically produced drug which, while not identical to physiologically produced oxytocin, is widely used to initiate or augment uterine contractions.

Oxytocin: a hormone produced in the hypothalamus and secreted by the pituitary gland which stimulates the myometrium (muscle wall) of the uterus to contract. Oxytocin also plays an important role in emotional bonding and can be released during peak moments such as orgasm and childbirth.

Pelvic floor: a sling-like network of muscles and connective tissue which spans the lower pelvis from front to back and supports the bladder, bowel, and reproductive organs.

Perimenopause: the life stage during which a person begins to experience hormonally controlled changes indicating the approach of menopause. Perimenopausal symptoms can affect every part of the body and mind, including vasomotor effects such as hot flashes and night sweats; mood changes; thinning and drying of vaginal tissues; and a reduction in bone mass.

Perimetrium: the outer layer of the uterus.

Peristalsis: involuntary, wavelike contractions of smooth muscle.

Pitocin: the brand name for a form of synthetically produced oxytocin, often used to initiate or augment uterine contractions, or to prevent or manage postpartum hemorrhage. *See also* Syntocinon.

Placenta: the organ created inside the uterus during pregnancy to nurture and sustain a fetus. Blood, oxygen, nutrients, and waste products are exchanged across the placenta in a sophisticated vascular system.

Postnatal or postpartum: after birth.

Postpartum hemorrhage: excessive blood loss in the immediate period after birth (a primary postpartum hemorrhage) or up to six weeks afterward (a secondary postpartum hemorrhage). Postpartum hemorrhage can be caused by retained pregnancy tissue (placenta and/or membranes), an overstimulated or atonic uterus, or trauma to the genital tract.

Preeclampsia: a potentially fatal disease of pregnancy, characterized by high blood pressure and protein in the urine.

Progesterone: a hormone produced by the corpus luteum in the nonpregnant female, or by the placenta during pregnancy. Progesterone thickens the endometrium and supports the pregnancy, and also contributes to overall health and well-being.

Prolapse: the displacement of an organ from its normal location within the body. The uterus can prolapse to varying degrees, descending into or even protruding from the vagina.

Pseudomenses: the "false" period experienced by female infants within the first week of life; a normal, fleeting discharge caused by fluctuating hormones.

Sepsis: a condition in which the body initiates an extreme systemic response to a localized infection, sometimes leading to tissue damage, organ failure, or death.

Smear or smear test: an investigation to detect cancerous or precancerous cells in the cervix. Sometimes known as a Pap smear, especially in America.

Speculum: an instrument used to hold open the vagina, enabling the operator to visualize clearly the vaginal walls and cervix.

Sterile: free from microorganisms such as bacteria or viruses.

Sterilization: the process of rendering a person infertile by surgical or other means.

Stillbirth: the birth of a fetus that has died in the womb; generally referring to a fetus over the legal age of viability.

Striae: stretch marks.

Stromal cells: a type of cell found throughout the body (for example, within the placenta) that is capable of differentiating and reorganizing into a number of different types of cells.

Syntocinon: the brand name for a form of synthetically produced oxytocin, often used to initiate or augment uterine contractions, or to prevent or manage postpartum hemorrhage. *See also* Pitocin.

Top surgery: a form of gender confirmation surgery in which a person's breast tissue is refashioned or reconstructed to match the sex with which they identify. For example, top surgery for a trans man could involve the removal of mammary tissue to create a flatter, more conventionally male-looking chest. Conversely, a trans woman may opt for breast augmentation, if wished.

Transgender: relating to or describing a person who does not identify with the gender assigned to them at birth.

Trophoblast: a blastocyst's outer cell layer, which then develops to form the placenta and chorion. The trophoblast must engage in a complex interaction with the lining of the uterus to establish a successful supply of blood and nutrients for the ongoing pregnancy.

Tubal ligation: a surgical method of sterilization in which the uterine tubes are cut and "tied" with sutures, cauterized, or clipped.

Uterine milk: a highly nutritious secretion produced by the glands within the endometrium in very early pregnancy to nourish the developing embryo.

Uterine tubes: slim tubes connecting the ovaries to the main body of the uterus. After ovulation, an egg travels along the tubes to await potential fertilization within the uterus. Sometimes called fallopian tubes.

Uterotonic: (adj.) having a contractile effect on the muscular body of the uterus, or (noun) a substance which has this effect.

Uterus: the strong, muscular organ which thickens and sheds its lining in menstruation, and gestates and expels a fetus in pregnancy and birth. The seat of all human life, and every person's first home. Also known as the womb.

Uterus didelphys: a condition in which there are two uteruses, each with its own cervix, and sometimes two vaginas as well.

Vagina: the fleshy internal passage which leads from the cervix to the exterior of the body.

Vein: a vessel carrying deoxygenated blood back to the heart.

Vulva: the external female genitalia; often referred to incorrectly as the vagina.

Womb: *see* uterus.

Zygote: a fertilized egg that results from union with a sperm.

CREDITS

Page vii: Sonya Renee Taylor, excerpt from *The Body Is Not an Apology: The Power of Radical Self-Love*, 2nd rev. ed. (Oakland, CA: Berrett-Koehler Publishers, 2021). Reprinted with kind permission of the author.

Page 1: Virginia Woolf, *The Waves*.

Page 13: Leila Chatti, excerpt from "Mubtadiyah," *Virginia Quarterly Review* 93, no. 3 (Summer 2017). Copyright © 2017, 2020 by Leila Chatti. Reprinted with permission of The Permissions Company, LLC on behalf of Copper Canyon Press, coppercanyonpress.org.

Page 66: Hollie McNish, excerpt from "Belly," in *Plum* (London: Picador, 2017). Reprinted with permission of Lewinsohn Literary Agency.

Page 159: Brenda Shaughnessy, excerpt from "Liquid Flesh," in *Our Andromeda* (Port Townsend, WA: Copper Canyon Press, 2012). Copyright © 2012 by Brenda Shaughnessy. Reprinted with permission of The Permissions Company, LLC on behalf of Copper Canyon Press, www.coppercanyonpress.org.

Page 170: Fortesa Latifi, excerpt from "chronic illness." Copyright 2016 by Fortesa Latifi. Reprinted with permission of the author.

Page 218: Phyllis Doyle Pepe, excerpt from "Hysterectomy." Reprinted with permission of the author.

Page 236: Saidiya Hartman, excerpt from *Wayward Lives, Beautiful Experiments: Intimate Histories of Riotous Black Girls, Troublesome Women, and Queer Radicals* (New York: W. W. Norton 2019).